AF410808

Functional Coatings

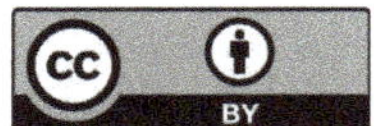

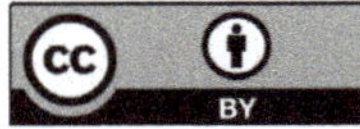

Functional Coatings: From Formulation in Solution to Applications on Surfaces and Interfaces

Functional Coatings: From Formulation in Solution to Applications on Surfaces and Interfaces

Selected Articles Published by MDPI

MDPI • Basel • Beijing • Wuhan • Barcelona • Belgrade • Manchester • Tokyo • Cluj • Tianjin

Contents

Preface to "Functional Coatings: From Formulation in Solution to Applications on Surfaces and Interfaces"

I started working with coatings about 20 years ago. I was interested in characterizing them rather than making them. Particularly, hydrophobic and liquid repellent coatings were interesting for me, and using high-speed imaging, I could work on observing droplet surface interactions in slow motion, such as contact angle dynamics, spreading kinetics, and droplet bounce. Later on, I realized that coatings are found everywhere including the food we eat, the clothes we wear, and the cars we drive, but we do not think much about them. Coatings always work behind the scenes to provide safety in numerous different ways, such as protection against wear and harsh friction, keep surfaces clean, prevent and delay corrosion, and block gas or toxic matter transport for security, to name a few. As of today, coatings have evolved tremendously. In general, multi-functional coatings can perform multiple tasks such as antimicrobial activity, liquid repellency, and sensorial response to various external stimuli. New technologies and formulations have been developed to increase the lifetime of coatings and to resist several harsh conditions such as acidic or alkaline environments, friction, abrasion and underwater exposure. As such, Coatings is a just platform to promote such recent experimental and theoretical advances in coatings science and technology. This particular collection gathers nine unique and well-executed works on coatings development and characterization. Each addresses important aspects of coatings from formulation in solution to interface adhesion and liquid repellency. The first three papers focus on water repellent polymer coatings, protective nanocomposite coatings for glass surfaces, and development of physical vapor deposited coatings to reduce adhesion during injection molding of plastic parts, respectively. The following set of three papers concentrates on the importance of polymer solution stability and the use of surfactants, preparation of novel waterborne hybrid polymer coating dispersions, and hydrolytic stability of adhesive interphases in composites. The last set of three papers emphasizes the durability of epoxy coatings applied over textured concrete surfaces, polymeric coatings of titanium implants via electrochemical processing, and analysis of fungal growth on coating-air interfaces in terms of coating pigmentation, cardinal direction, and the angle of exposure. In summary, these nine selected papers highlight active and important features in coating science and technology and show us that exciting advances both in coating fabrication and in testing are occurring. However, coatings fail one way or the other. For this reason, we need to work harder to develop and formulate new coating technologies to render them more sustainable, long lasting, resistant, functional, and responsive.

Ilker S. Bayer

Article

Preparation of Hierarchically Structured Polystyrene Surfaces with Superhydrophobic Properties by Plasma-Assisted Fluorination

Martin Minařík [1,2], Erik Wrzecionko [1,2], Antonín Minařík [1,2], Ondřej Grulich [1,2], Petr Smolka [1,2], Lenka Musilová [1,2], Ita Junkar [3], Gregor Primc [3], Barbora Ptošková [1,2], Miran Mozetič [3] and Aleš Mráček [1,2,*]

[1]   Department of Physics and Materials Engineering, Faculty of Technology, Tomas Bata University in Zlín, Vavrečkova 275, 760 01 Zlín, Czech Republic; mminarik@utb.cz (M.M.); wrzecionko@utb.cz (E.W.); minarik@utb.cz (A.M.); ondrej.grulich85@gmail.com (O.G.); smolka@utb.cz (P.S.); lmusilova@utb.cz (L.M.); b_ptoskova@utb.cz (B.P.)

[2]   Centre of Polymer Systems, Tomas Bata University in Zlín, Třída Tomáše Bati 5678, 76001 Zlín, Czech Republic

[3]   Jožef Stefan Institute, Jamova cesta 39, 1000 Ljubljana, Slovenia; ita.junkar@ijs.si (I.J.); gregor.primc@ijs.si (G.P.); miran.mozetic@guest.arnes.si (M.M.)

*   Correspondence: mracek@utb.cz; Tel.: +420-733-690-668

Received: 17 February 2019; Accepted: 16 March 2019; Published: 20 March 2019

Abstract: The nanotexturing of microstructured polystyrene surfaces through CF_4 plasma chemical fluorination is presented in this study. It is demonstrated that the parameters of a surface micropore-generation process, together with the setup of subsequent plasma-chemical modifications, allows for the creation of a long-term (weeks) surface-stable micro- and nanotexture with high hydrophobicity (water contact angle >150°). Surface micropores were generated initially via the time-sequenced dosing of mixed solvents onto a polystyrene surface (Petri dish) in a spin-coater. In the second step, tetrafluoromethane (CF_4) plasma fluorination was used for the generation of a specific surface nanotexture and the modulation of the surface chemical composition. Experimental results of microscopic, goniometric, and spectroscopic measurements have shown that a single combination of phase separation methods and plasma processes enables the facile preparation of a wide spectrum of hierarchically structured surfaces differing in their wetting properties and application potentials.

Keywords: surface pores; polystyrene; nanotexture; plasma; superhydrophobic

---

1. Introduction

Hierarchically structured surfaces with well-defined textures play very significant roles in sophisticated applications in sensors [1], photonics [2], tissue engineering [3], and superhydrophobic materials [4–12]. Generally, these surfaces are prepared in several steps combining mechanical, laser machining, physicochemical, and plasma technologies [13–16]. The so-called "breath figures" is one of the very popular and promising approaches [17–19]. In principle, the polymer surface is swollen by a "good" volatile solvent. This process is aimed at a defined humidity and temperature. The temperature of the swollen polymer layer decreases with the solvent evaporation, and the water vapor condensates on it. These water drops can ideally create hexagonal organized structures called "breath figures". The droplets do not coalesce because of the Marangoni convections [20] in the droplets or the precipitation of polymers onto the water droplet interfaces [21]. The "breath figures" method has many variations and technological modifications, and some of them can lead to the formation of very impressive honeycomb-like structures, not only in the thin surface layer, but porous structures in bulk can also be created [4,13,21,22].

Recently, other methods based on similar principles of mixed (good and pure solvents) solutions have been published [23]. This approach is based on the time-sequenced dispensing of a mixture of good and poor solvents on the rotating polystyrene surface. The phase separation—a process that occurs during application to the surface of the substrate—may be caused by temperature change, poor solvents [24,25], chemical reactions [26], or shear deformation [27]. The same process, published in recent work [23], was used in this paper for the preparation of the microstructured porous surfaces.

The excellent hydrophobic properties of a surface are not achieved solely by chemical compositions. The hierarchical geometric structure of the surface is necessary for the achievement of a water contact angle above 160° [28–30]. We can find many theoretical models (e.g., Wenzel or Cassie–Baxter models) describing the influence of the structure and the chemical composition of the surface on wetting [31–39]. Surface functionalization by the incorporation of non-polar groups is very often used for improving hydrophobicity. There are numerous techniques for the chemical modification of surface properties, but one of them is predominant nowadays. Plasma surface modifications are very effective tools for chemical treatment, etching, thin film deposition, and nanofabrication [40–50]. Inductively coupled plasma with CF_4 or C_4F_8 as process gases can be used to render polystyrene surfaces superhydrophobic [51,52]. The man-made superhydrophobic surfaces, however, are subject to aging, and with time, they lose their water-repellent properties [52]. These undesirable processes can be caused by the reorganization of chemical groups from the surface to the bulk and also by the water vapor (humidity) [53,54].

This paper is related to our recent research dealing with the creation of microstructured porous surfaces of polystyrene [23]. Those surfaces were originally prepared for bio-applications, with a water contact angle below 115°. In this work, CF_4 plasma was used for nanostructure generation on a polystyrene microstructured surface that was recently prepared. As can be seen, hierarchically structured polystyrene with non-polar chemical groups on the surface (CF and CF_2) exhibited superhydrophobic properties (water contact angle over 155°), and these properties were stable over several weeks.

2. Materials and Methods

2.1. Materials

Polystyrene (PS) Petri dishes with a diameter of 3.4 cm, radiation-sterilized, free from pyrogens, and with DNA/RNA for cell cultivation (TPP Techno Plastic Products AG, Trasadingen, Switzerland), were used as substrates. Tetrahydrofuran—HPLC grade (THF) and 2-ethoxyethanol p.a. (ETH), both from Sigma–Aldrich Ltd., (St. Louis, MO, USA) were used for surface microstructure production on the PS substrates [23]. Ar gas (purity 99.999%, Messer Bad Soden am Taunus, Germany) and CF_4 gas (purity 99.7%, Air Liquide, Paris, France) were used.

The surfaces were modified with spin-coating. The solvent mixture was deposited onto the surface of the PS dishes with a specially constructed dosing device (Figure 1) rotating at 2200 rpm. The dosing of solvents was carried out by using a syringe placed 30 mm above the center of the rotating substrate. Each time, 0.4, 1.0, or 1.6 mL of a mixed solvent, divided into two, five, or eight consecutive doses of about 200 µL, was deposited in 5 s intervals on the surface of the PS dishes. After the last dose, the sample was left to rotate for another 120 s. Unless otherwise stated, all the experiments were performed at a temperature of 295 K (substrate, solutions, and surrounding atmosphere) or a temperature of 303 K for solutions. Besides that, the air humidity was monitored and kept at 50% ± 2%, as described in our previous work [23].

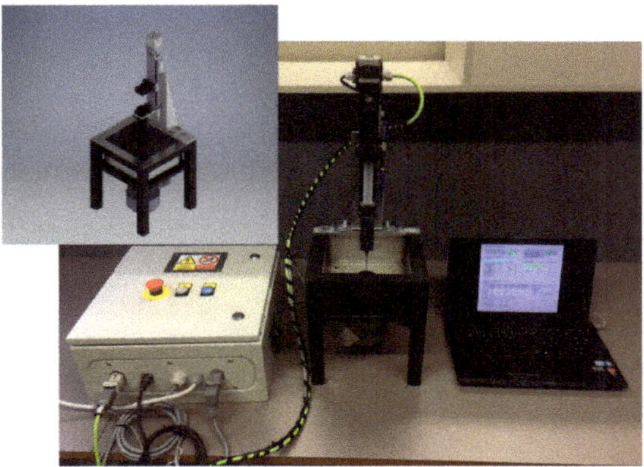

Figure 1. The dosing device constructed for the modification of the polymer surface topography (TSSC) by phase separation at rotation. The device consists of the sample carrier on the rotor, the dosing unit, and the control electronics.

A specially constructed device (Figure 1), a time-sequenced phase-separation chamber (TSSC, home made), is proposed for the generation of micro- and nanoporous polymer systems at rotation. The TSSC allows for the control of all of the process parameters of the previously described time-sequenced phase separations during rotation [23].

2.2. Plasma Surface Modification

Plasma treatment was performed in inductively coupled plasma in a discharge borosilicate glass tube with a full length of 80 cm and a 4 cm inner diameter. The gas pressure of 70 Pa in the glow-chamber was kept constant. Plasma was created by the radiofrequency (RF) generator Caesar 1312 (Advanced Energy, Warstein-Belecke, Germany) (Advanced Energy) coupled with a coil with six turns via a matching network. The matching network consisted of two vacuum-tunable capacitors. The generator operated at a standard frequency of 13.56 MHz and an adjustable nominal power of up to 1200 W. The matching system was optimized for H mode (forward power over ~500 W and low reflected power) [46], but E mode (less than 500 W of forward power) was used for sample modification due to expected degradation. The RF power was varied, as is described further in the text. The plasma processing time was 2 s for Ar activation and 240 s for CF_4 modification. The processing was identical for both the first and second plasma treatment steps.

2.3. Scanning Electron Microscopy (SEM)

Modified samples were analyzed with the JEOL JSM 6060 LV (JEOL USA, Inc., Peabody, MA, USA) and Phenom Pro (Phenom-World BV, Eindhoven, The Netherlands) scanning electron microscopes (SEM). Samples were observed at an acceleration voltage ranging from 1 to 5 kV in the backscattered electron mode at a magnification ranging from 2000× to 50,000×. Measurements were carried out on samples without prior metallization, using a special sample holder for the Phenom Pro, or with a carbon coating for the JEOL JSM 6060 LV in low vacuum mode.

2.4. Atomic Force Microscopy (AFM)

The surface topography was characterized by an atomic force microscope (AFM). The models Dimension ICON (Bruker, Santa Barbara, CA, USA) and NTEGRA-Prima (NT–MDT, Spectrum Instruments, Moscow, Russia) were used. Measurements were performed at a scan speed ranging from 0.3 to 0.7 Hz, with a resolution of 512 × 512 pixels in tapping mode at room temperature under an air atmosphere. NSG01 (AppNano, Inc., Santa Clara, CA, USA) silicone probes were used.

2.5. Profilometry

Changes in the surface roughness (R_a) were characterized by a contact profilometer (DektaXT, Bruker, Billerica, MA, USA). A diamond tip with a radius of curvature of 2 microns was used. The evaluation of the surface roughness was performed according to the ASME B46.1 standard [55]. The mean R_a values were determined from 10 individual measurements at various locations on the three samples.

2.6. Distribution of Micropores

The distribution of the pore areas was obtained by image analysis using ImageJ 1.5 software (Wayne Rasband, National Institutes of Health, Bethesda, MD, USA). The size of the individual micropores was determined from the thresholded SEM images.

2.7. Contact Angle Measurement

The sliding, advancing, receding, and static contact angles of water (θ) on the PS surface were characterized by the Drop Shape Analyzer—DSA 30 (Krüss GmbH, Hamburg, Germany). Measurements were done at room temperature (298 ± 1 K) with 50% humidity. A drop of 3 μL (for the static contact angle) or 10 μL (for the sliding, advancing, and receding contact angle) was deposited onto the measured surface. Ultrapure water with a resistance of 18.2 MΩ cm was used for the measurement. All measurements were repeated 10 times, and the mean values and standard deviations are reported.

2.8. X-ray Photoelectron Spectroscopy (XPS)

Samples were analyzed with the TFA XPS instrument (Physical Electronics, Lake Drive East Chanhassen, MN, USA). The base pressure in the chamber was about 6×10^{-8} Pa. The samples were excited with X-rays over a 400 μm spot area, with a monochromatic Al Kα with a radiation energy of 1486.6 eV and a linewidth of 1.2 eV. The photoelectrons were detected with a hemispherical analyzer positioned at an angle of 45° with respect to the perpendicular of the sample surface. Survey-scan spectra were acquired at a pass energy of 187.85 eV and an energy step of 0.4 eV, while individual high-resolution spectra for O $1s$ and F $1s$ were taken at a pass energy of 23.5 eV and an energy step of 0.1 eV, and at 11.75 eV and 0.05 eV for C $1s$, respectively. An electron gun was used for surface neutralization. All spectra (not containing F) were referenced to the main C1 peak of the carbon atoms, which was assigned a value of 284.8 eV. The spectra containing F were shifted to 291.8 eV (CF_2 binding of the carbon atom). The concentrations of the elements and the concentrations of the different chemical states of the carbon atoms in the C1 peaks were determined using the MultiPak v7.3.1 software from Physical Electronics. Carbon C1 peaks were fitted with symmetrical Gauss–Lorentz functions and the Shirley-type background subtraction was used.

3. Results and Discussion

For this study, PS samples with various surface structures were prepared according to the process described in our previous work [23]. It was found that the deposition of the heated solvent mixture (303 K) led to the formation of nanopores at the micropore edges (Figure 2 and Figure 4). These secondary pores are not attributed to higher hydrophobicities, as demonstrated by the water contact angle values in Figure 4c ($115° \pm 2°$) and Figure 4e ($107° \pm 2°$). To further increase the hydrophobicity, homogenous surface nanotextures have to be created, similar to natural materials [28], and CF_x functional groups have to be introduced onto the surface.

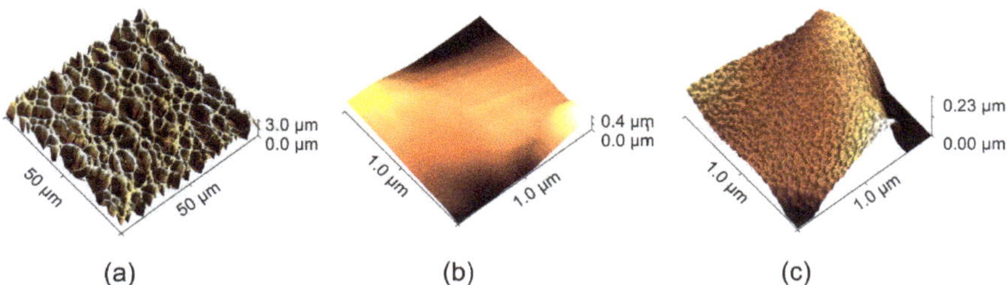

(a)   (b)   (c)

Figure 2. Surface micro- and nanotextures at the micropore border generated by inductively coupled Ar and CF_4 plasma: (**a**) microporous polystyrene (PS) with a smooth border; (**b**) detail of the smooth border before plasma treatment; (**c**) detail of the smooth border after plasma treatment. All images represent atomic force microscope (AFM) micrographs. The smooth borders were prepared with five doses of 1.5 THF:8.5 ETH at 295 K. The power of the plasma reactor was 300 W.

Within this study, the plasma treatment process was optimized for PS microtextured surface modification [23], in order to create secondary surface corrugation (Figure 2 and Figure 4). Plasma treatment parameters, such as treatment time, plasma power, and gas flow, had to be optimized. Some of the parameters will be discussed in the context of primary pore generation with the time-sequenced phase separation technique.

As can be seen from the data in Figure 3, the most hydrophobic surface was reached at 300 W plasma power and a contact angle of 146° ± 2°. Lower plasma power (250 W) resulted in less intense surface modification and a final contact angle of 139° ± 1°. Higher plasma power (350 W), on the other hand, caused unwanted surface degradation and the contact angle reached 143° ± 1°.

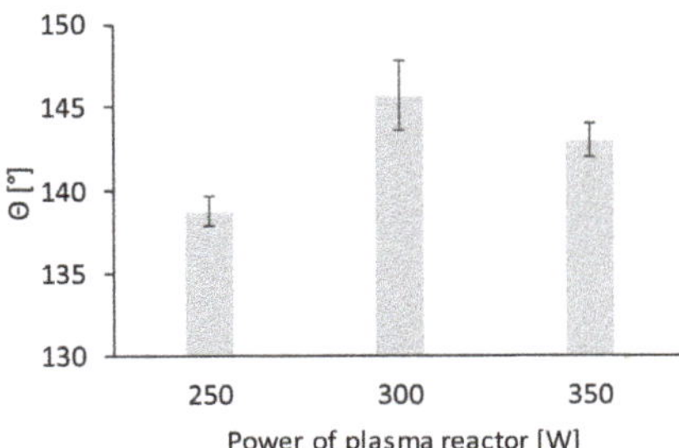

Figure 3. Water contact angle vs. plasma power on the PS surface with smooth pore borders with one cycle of plasma treatment. The smooth borders were prepared using five doses in 1.5 THF:8.5 ETH at 295 K.

The experiments revealed that in order to achieve homogenous surface corrugation on the PS pore boundaries (Figure 4b,d,f) and to stabilize the surface modifications, it is necessary to set the optimal plasma power (300 W) and to repeat the plasma treatment twice. The data from Figure 4 suggest that the specific nanotexture can be generated, regardless of the initial surface topography. With the smooth surface, the water contact rises by an angle of 32° (Figure 4a,b), while at the microporous surface, it rises by 43° (Figure 4c–f). This observation is related to the well-known effect of the combined micro- and nanotexture upon surface wetting [28]. These results could possibly indicate that the process of primary pore generation with phase separation does not affect the wetting characteristics of the plasma-treated surfaces. However, these appearances are deceptive, as the absolute values of the water contact sliding angles were 5° ± 1° (Figure 4d) and 8° ± 1° (Figure 4f). The corresponding advancing and receding water contact angles were 158° ± 1° and 156° ± 1°, respectively for Figure 4d and 151° ± 1° and 148° ± 1°, respectively, for Figure 4f.

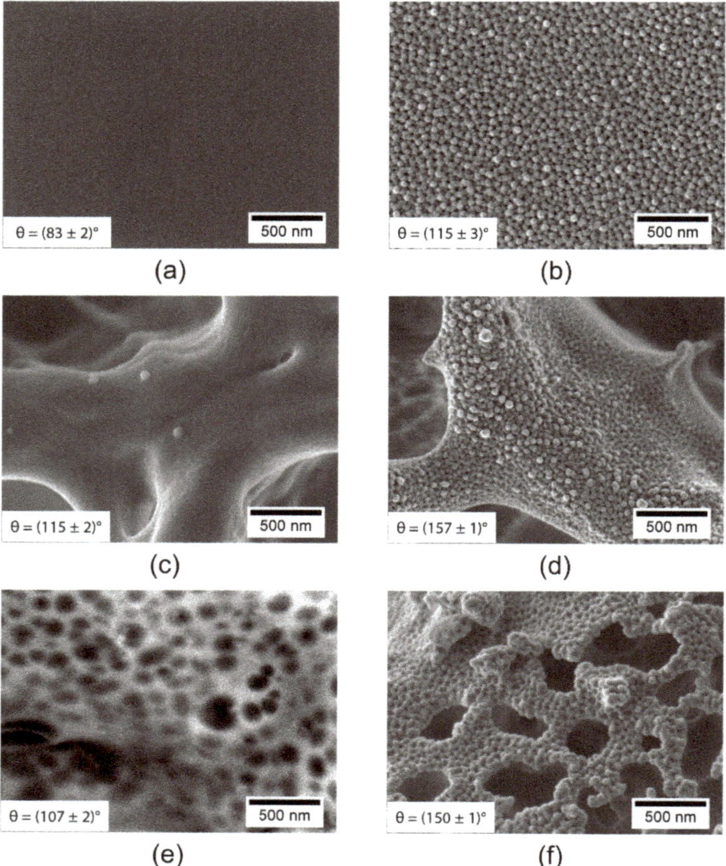

Figure 4. The effect of plasma etching on different types of PS substrate and the corresponding water contact angle: (**a**) flat PS; (**b**) CF_4 plasma-treated flat PS; (**c**) porous PS with a smooth border; (**d**) CF_4 plasma-treated porous PS with a smooth border; (**e**) porous PS with a porous border; (**f**) CF_4 plasma-treated porous PS with a porous border. The images represent SEM micrographs. The smooth borders were prepared using five doses of 1.5 THF:8.5 ETH at 295 K. The porous borders were prepared using five doses of 2.5 THF:7.5 ETH at 303 K. The power of the plasma reactor was 300 W.

Two distinct surface morphologies, differing in pore border appearance were used for plasma treatment: (1) a smooth border (Figure 4c) including five doses of the solvent mixture (THF:ETH in a volume ratio of 1.5:8.5) deposited in five second intervals at 295 K, and (2) a porous border (Figure 4e) including five doses of the solvent mixture (THF:ETH, in a volume ratio of 2.5:7.5) deposited in five second intervals at 303 K.

The stability of the plasma treatment was observed by means of the water contact angle measurements. The data were recorded on the first, second, third, and fourth days after plasma treatment (Figure 5). As expected, the untreated samples remained stable during the whole time period. The plasma-treated samples stayed stable up to the second day, and then a slight decrease in contact angle value was observed. After 14 days, the water contact angle remained in the range of 140°–152°, depending on the initial surface microstructure.

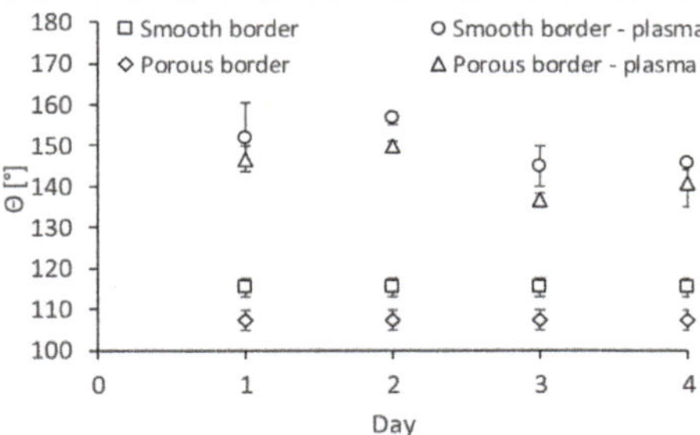

Figure 5. Water contact angle vs. time on the plasma-treated samples of PS substrates with smooth borders and porous borders and two cycles of plasma treatment. The smooth borders were prepared by five doses of 1.5 THF:8.5 ETH, at 295 K. The porous borders were prepared with five doses of 2.5 THF:7.5 ETH at 303 K. The power of the plasma reactor was 300 W.

The initial experiments revealed that the two-fold plasma treatment resulted in the elimination of the bonded oxygen from the surface, and thus, the water contact angle rose (Table 1). This corresponded with the accelerated reorganization of the surface functional groups in the second cycle of the plasma treatment. Otherwise, this change would have occurred spontaneously in the following days after the first plasma treatment cycle [53]. This hypothesis is further supported by the increase in the water contact angle value in the second and third day after the one-cycle plasma treatment.

Table 1. Surface composition of the microporous PS with the smooth border and the corresponding water contact angles after the repeated plasma treatment. The smooth borders were prepared with five doses of 1.5 THF:8.5 ETH at 295 K. The power of the plasma reactor was 300 W.

Porous PS (Smooth Border) (XPS Characterization Immediately after Processing)	XPS Elemental Composition ($\pm$ 0.5%)			θ [°]
	C (%)	F (%)	O (%)	
First plasma treatment	38.4	56.7	5.0	142 $\pm$ 3
Second plasma treatment	43.0	56.3	0.7	157 $\pm$ 1

The sample surface of the chemical composition and the type of chemical bonds did not change, even after 14 days, as can be seen from the comparison in Table 1 (second row) and Table 2 (fourth row). The data in Table 2 show that the highest bonded fluorine concentration was observed in microporous smooth-border samples and porous-border samples. C–CF, C–F, and CF$_2$, and in some cases, CF$_3$ dominated over the C–C bonds, which, in turn, dominated the smooth plasma-treated PS surface.

Table 2. XPS characterization of plasma-untreated and treated PS substrates 14 days from processing. The smooth borders were prepared by five doses of 1.5 THF:8.5 ETH at 295 K. Porous borders were prepared with five doses of 2.5 THF:7.5 ETH at 303 K. The power of the plasma reactor was 300 W.

Sample Type (XPS Characterization 14 Days after Processing)	XPS Elemental Composition ($\pm$ 0.5%)			Type of Chemical Bonds				
	C (%)	F (%)	O (%)	C–C (%)	C–CF (%)	C–F (%)	CF$_2$ (%)	CF$_3$ (%)
Flat PS	96.6	–	2.6	100.0	–	–	–	–
Plasma—Flat PS	44.8	54.6	0.6	23.7	12.1	14.1	37.2	13.0
Porous PS (smooth border)	94.6	–	3.5	100.0	–	–	–	–
Plasma—Porous PS (smooth border)	42.9	56.6	0.5	15.6	15.0	22.2	38.2	9.1
Porous PS (porous border)	97.0	–	2.5	100.0	–	–	–	–
Plasma—Porous PS (porous border)	41.7	58.3	1.0	14.2	11.6	24.3	37.7	12.3

The discussed smooth-border and porous-border surfaces have distinct pore sizes and size distributions (Figures 6 and 7). Still, stemming from the inserts in Figure 6, the water contact angle before the plasma treatment was almost identical on all surfaces. Only after two cycles of plasma treatment could differences be observed. The highest value of the contact angle (157°) was achieved at the surface, with an intermediate pore size and roughness of 117 μm. The surfaces with the smallest and largest pores had water contact angles of 145° and 146°, respectively. To achieve maximum hydrophobicity, it was necessary to adjust the pore size (Figure 6d) and depth (Figure 6e).

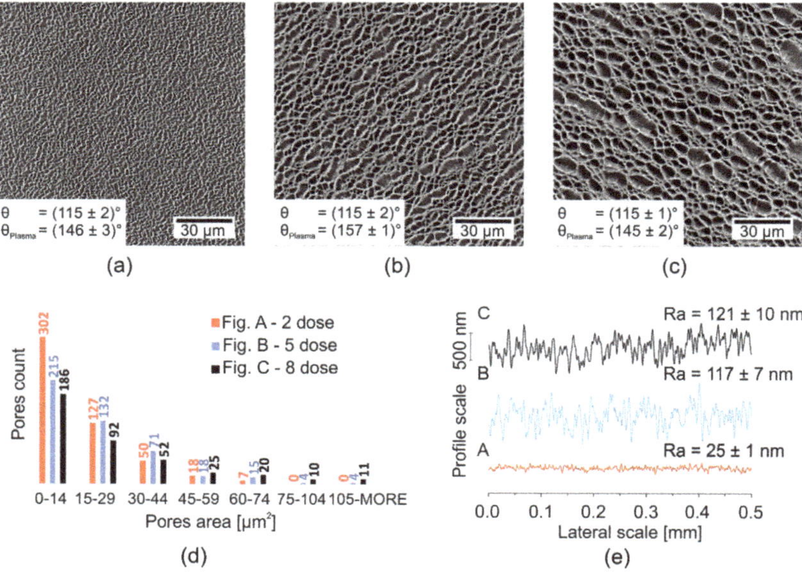

Figure 6. The effect of PS surface microtopography on the water contact angle values after plasma modification. The smooth borders of the microtextured surfaces were prepared from (**a**) two, (**b**) five, and (**c**) eight doses used in the time-sequenced phase separation process [23]. Pore size distribution (**d**), and surface profiles (**e**). SEM micrographs (**a–c**). The smooth borders were prepared with two, five, or eight doses of 1.5 THF:8.5 ETH at 295 K. The power of the plasma reactor was 300 W.

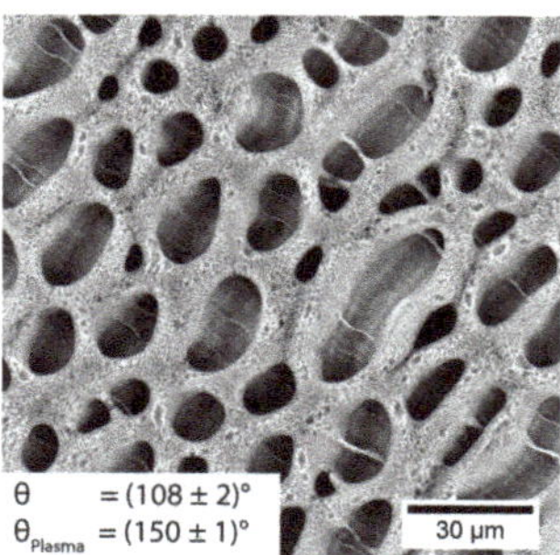

Figure 7. SEM micrograph of microtextured PS with a porous border. Water contact angles prior to and after plasma modification. Porous-borders were prepared by five doses of 2.5 THF:7.5 ETH at 303 K. The power of plasma reactor was 300 W.

Other situations occurred, with secondary nanopores presented on the micropore boundaries (Figure 7). Such kinds of surfaces have high contact angle values, although small amounts of large pores are present on the surface, and the roughness is relatively very high (R_a = 634 ± 57 μm).

4. Conclusions

The potential of the plasma hydrophobization of micro- and nanoporous PS substrates was studied. It was found that the single combination of phase-separation methods and CF_4 plasma enabled the quick preparation of various types of hierarchically structured surfaces with superhydrophobic properties. It was also found that the hydrophobic behavior is dictated not only by the pore size and thickness but also by the interface separating the individual pores generated in the phase separation process. With the appropriate plasma power and repeated exposure to the plasma, highly hydrophobic surfaces with specific surface nanotextures can be prepared. Such induced changes, both topographic and chemical, are very stable, and the treated surfaces are not subject to aging within a time period of several weeks.

Author Contributions: Conceptualization, A.M. (Aleš Mráček) and A.M. (Antonín Minařík); Methodology, M.M. (Martin Minařík), A.M. (Antonín Minařík), O.G., E.W. and A.M. (Aleš Mráček); Validation, M.M. (Martin Minařík), L.M., A.M. (Antonín Minařík) and A.M. (Aleš Mráček); Formal Analysis, A.M. (Antonín Minařík), B.P., M.M. (Martin Minařík) and O.G.; Investigation, M.M. (Martin Minařík), A.M. (Antonín Minařík), P.S., E.W. and A.M. (Aleš Mráček); Resources, A.M. (Aleš Mráček) and A.M. (Antonín Minařík); Writing–Original Draft Preparation, A.M. (Aleš Mráček) and A.M. (Antonín Minařík); Writing–Review and Editing, A.M. (Aleš Mráček) A.M. (Antonín Minařík) and P.S.; Visualization, M.M. (Martin Minařík), E.W., O.G. and A.M.; supervision, A.M. (Aleš Mráček), M.M. (Miran Mozetič), A.M. (Antonín Minařík), I.J. and G.P.

Funding: The research was funded by the Ministry of Education, Youth and Sports of the Czech Republic—Program NPU I (LO1504) and the European Regional Development Fund (No. CZ.1.05/2.1.00/19.0409) as well as by TBU (Nos. IGA/FT/2017/011, IGA/FT/2018/011, and IGA/FT/2019/012) funded from the resources for specific university research.

Conflicts of Interest: The authors declare no conflict of interest.

References

1.   Perez, J.M.; O'Loughin, T.; Simeone, F.J.; Weissleder, R.; Josephson, L. DNA-based magnetic nanoparticle assembly acts as a magnetic relaxation nanoswitch allowing screening of DNA-cleaving agents. *J. Am. Chem. Soc.* **2002**, *124*, 2856–2857. [CrossRef]

2.   Imada, M.; Noda, S.; Chutinan, A.; Tokuda, T.; Murata, M.; Sasaki, G. Coherent two-dimensional lasing action in surface-emitting laser with triangular-lattice photonic crystal structure. *Appl. Phys. Lett.* **1999**, *75*, 316–318. [CrossRef]

3.   Shastri, V.P.; Martin, I.; Langer, R. Macroporous polymer foams by hydrocarbon templating. *Proc. Natl. Acad. Sci. USA* **2000**, *97*, 1970–1975. [CrossRef] [PubMed]

4.   Brown, P.S.; Talbot, E.L.; Wood, T.J.; Bain, C.D.; Badyal, J.P.S. Superhydrophobic Hierarchical Honeycomb Surfaces. *Langmuir* **2012**, *28*, 13712–13719. [CrossRef]

5.   Wasser, L.; Vacche, D.S.; Karasu, F.; Müller, L.; Castellino, M.; Vitale, A.; Bongiovanni, R.; Leterrier, Y. Bio-inspired fluorine-free self-cleaning polymer coatings. *Coatings* **2018**, *8*, 436. [CrossRef]

6.   Huang, Z.; Xu, W.; Wang, Y.; Wang, H.; Zhang, R.; Song, X.; Li, J. One-step preparation of durable super-hydrophobic MSR/SiO$_2$ coatings by suspension air spraying. *Micromachines* **2018**, *9*, 677. [CrossRef]

7.   Jia, S.; Deng, S.; Luo, S.; Qing, Y.; Yan, N.; Wu, Y. Texturing commercial epoxy with hierarchical and porous structure for robust superhydrophobic coatings. *Appl. Surf. Sci.* **2019**, *466*, 84–91. [CrossRef]

8.   Ishizaki, T.; Kumagai, S.; Tsunakawa, M.; Furukawa, T.; Nakamura, K. Ultrafast fabrication of superhydrophobic surfaces on engineering light metals by single-step immersion process. *Mater. Lett.* **2017**, *193*, 42–45. [CrossRef]

9.   Boinovich, L.B.; Emelyanenko, A.M.; Modestov, A.D.; Domantovsky, A.G.; Emelyanenko, K.A. Not simply repel water: The diversified nature of corrosion protection by superhydrophobic coatings. *Mendeleev Commun.* **2017**, *27*, 254–256. [CrossRef]

10. Kadlečková, M.; Minařík, A.; Smolka, P.; Mráček, A.; Wrzecionko, E.; Novák, L.; Musilová, L.; Gajdošík, R. Preparation of textured surfaces on aluminum-alloy substrates. *Materials* **2018**, *12*, 109. [CrossRef]

11. Wang, S.; Liu, K.; Yao, X.; Jiang, L. Bioinspired surfaces with superwettability: New insight on theory, design, and applications. *Chem. Rev.* **2015**, *115*, 8230–8293. [CrossRef] [PubMed]

12. Su, B.; Tian, Y.; Jiang, L. Bioinspired interfaces with superwettability: From materials to chemistry. *J. Am. Chem. Soc.* **2016**, *138*, 1727–1748. [CrossRef] [PubMed]

13. Widawski, G.; Rawiso, M.; François, B. Self-organized honeycomb morphology of star-polymer polystyrene film. *Nature* **1994**, *369*, 387–389. [CrossRef]

14. Beysens, D.; Knobler, C.M. Growth of breath figures. *Phys. Rev. Lett.* **1986**, *57*, 1433–1436. [CrossRef] [PubMed]

15. Ahmmed, K.M.T.; Mafi, R.; Kietzig, A.M. Colored poly (vinyl chloride) by femtosecond laser machining. *Ind. Eng. Chem. Res.* **2018**, *57*, 6161–6170. [CrossRef]

16. Ahmmed, K.M.T.; Patience, C.; Kietzig, A.M. Internal and external flow over laser-textured superhydrophobic polytetrafluoroethylene (PTFE). *ACS Appl. Mater. Interfaces* **2016**, *8*, 27411–27419. [CrossRef] [PubMed]

17. Aitken, J. Breath figures. *Nature* **1911**, *86*, 516–517. [CrossRef]

18. Rayleigh, L. Breath figures. *Nature* **1911**, *86*, 416–417. [CrossRef]

19. Rayleigh, L. Breath figures. *Nature* **1912**, *90*, 436–438. [CrossRef]

20. Bormashenko, E.; Balter, S.; Pogreb, R.; Bormashenko, Y.; Gendelman, O.; Aurbach, D. On the mechanism of patterning in rapidly evaporated polymer solutions: Is temperature-gradient-driven marangoni instability responsible for the large-scale patterning? *J. Colloid Interface Sci.* **2010**, *343*, 602–607. [CrossRef]

21. Stenzel-Rosenbaum, M.H.; Davis, T.P.; Fane, A.G.; Chen, V. Porous polymer films and honeycomb structures made by the selforganization of well-defined macromolecular structures created by living radical polymerization techniques. *Angew. Chem. Int. Ed.* **2001**, *40*, 3428–3432. [CrossRef]

22. Tian, Y.; Jiao, Q.; Ding, H.; Shi, Y.; Liu, B. The formation of honeycomb structure in polyphenylene oxide films. *Polymer* **2006**, *47*, 3866. [CrossRef]

23. Wrzecionko, E.; Minařík, A.; Smolka, P.; Minařík, M.; Humpolíček, P.; Rejmontová, P.; Mráček, A.; Minaříková, M.; Gřundělová, L. Variations of polymer porous surface structures via the time-sequenced dosing of mixed solvents. *ACS Appl. Mater. Inter.* **2017**, *9*, 6472–6481. [CrossRef] [PubMed]

24. DeRosa, M.; Hong, Y.; Faris, R.; Rao, H. Microtextured polystyrene surfaces for three- dimensional cell culture made by a simple solvent treatment method. *J. Appl. Polym. Sci.* **2014**, *131*, 40181–40190. [CrossRef]

25. Samuel, A.; Umapathy, S.; Ramakrishnan, S. Functionalized and postfunctionalizable porous polymeric films through evaporation-induced phase separation using mixed solvents. *ACS Appl. Mater. Inter.* **2011**, *3*, 3293–3299. [CrossRef]

26. Li, W.; Ryan, A.J.; Meier, I.K. Morphology development via reaction-induced phase separation in flexible polyurethane foam. *Macromolecules* **2002**, *35*, 5034–5042. [CrossRef]

27. Matsuzaka, K.; Jinnai, H.; Koga, T.; Hashimoto, T. Effect of oscillatory shear deformation on demixing processes of polymer blends. *Macromolecules* **1997**, *30*, 1146–1152. [CrossRef]

28. Bhushan, B.; Jung, Y.C. Natural and biomimetic artificial surfaces for superhydrophobicity, self-cleaning, low adhesion, and drag reduction. *Prog. Mater. Sci.* **2011**, *56*, 1–108. [CrossRef]

29. Ma, M.; Hill, R.M. Superhydrophobic surfaces. *Curr. Opin. Colloid Interface Sci.* **2006**, *11*, 193–202. [CrossRef]

30. Xiu, Y.; Zhu, L.; Hess, D.W.; Wong, C.P. Hierarchical silicon etched structures for controlled hydrophobicity/superhydrophobicity. *Nano Lett.* **2007**, *7*, 3388–3393. [CrossRef]

31. Wenzel, R.N. Resistance of solid surfaces to wetting by water. *Ind. Eng. Chem.* **1936**, *28*, 988–994. [CrossRef]

32. Baxter, S.; Cassie, A.B.D. The water repellency of fabrics and a new water repellency test. *J. Text. Inst. Trans.* **1945**, *36*, T67–T90. [CrossRef]

33. Cassie, A.B.D.; Baxter, S. Wettability of porous surfaces. *Trans. Faraday Soc.* **1944**, *40*, 546–551. [CrossRef]

34. Cassie, A.B.D.; Baxter, S. Large contact angles of plant and animal surfaces. *Nature* **1945**, *155*, 21–22. [CrossRef]

35. Cassie, A.B.D. Contact angles. *Discuss. Faraday Soc.* **1948**, *3*, 11–16. [CrossRef]

36. Nosonovsky, M.; Bhushan, B. Roughness optimization for biomimetic superhydrophobic surfaces. *Microsyst. Technol.* **2005**, *11*, 535–549. [CrossRef]

37. Jung, Y.C.; Bhushan, B. Contact angle, adhesion, and friction properties of micro- and nanopatterned polymers for superhydrophobicity. *Nanotechnology* **2006**, *17*, 4970–4980. [CrossRef]

38. Milne, A.J.B.; Amirfazli, A. The Cassie equation: How it is mean to be used. *Adv. Colloid Interface Sci.* **2012**, *170*, 48–55. [CrossRef] [PubMed]

39. Bormashenko, E. Progress in understanding wetting transitions on rough surfaces. *Adv. Colloid Interface Sci.* **2015**, *222*, 92–103. [CrossRef] [PubMed]

40. Mozetič, M.; Zalar, A.; Panjan, P.; Bele, M.; Pejovnik, S.; Grmek, R. A method of studying carbon particle distribution in paint films. *Thin Solid Films* **2000**, *376*, 5–8. [CrossRef]

41. Gunde, M.K.; Kunaver, M.; Mozetič, M.; Pelicon, P.; Simčič, J.; Budnar, M.; Bele, M. Microstructure analysis of metal-effect coatings. *Surf. Coat. Int. B Coat. Trans.* **2002**, *85*, 115–121. [CrossRef]

42. Kunaver, M.; Gunde, M.K.; Mozetič, M.; Hrovat, A. The degree of dispersion of pigments in powder coatings. *Dyes Pigment.* **2003**, *57*, 235–243. [CrossRef]

43. Mozetič, M. Controlled oxidation of organic compounds in oxygen plasma. *Vacuum* **2003**, *71*, 237–240. [CrossRef]

44. Kunaver, M.; Mozetič, M.; Gunde, M.K. Selective plasma etching of powder coatings. *Thin Solid Films* **2004**, *459*, 115–117. [CrossRef]

45. Drenik, A.; Vesel, A.; Mozetič, M. Controlled carbon deposit removal by oxygen radicals. *J. Nucl. Mater.* **2009**, *386–388*, 893–895. [CrossRef]

46. Mozetič, M. Surface modification of materials using an extremely non-equilibrium oxygen plasma. *Mater. Tehnol.* **2010**, *44*, 165–171.

47. Drenik, U.; Cvelbar, K.; Ostrikov, M.; Mozetič, M. Catalytic probes with nanostructured surface for gas/discharge diagnostics: A study of a probe signal behaviour. *J. Phys. D Appl. Phys.* **2008**, *41*, 115201. [CrossRef]

48. Ricard, A.; Gaillard, M.; Monna, V.; Vesel, A.; Mozetič, M. Excited species in H_2, N_2, O_2 microwave flowing discharges and post-discharges. *Surf. Coat. Technol.* **2001**, *142–144*, 333–336. [CrossRef]

49. Babič, D.; Poberaj, I.; Mozetič, M. Fiber optic catalytic probe for weakly ionized oxygen plasma characterization. *Rev. Sci. Instrum.* **2001**, *72*, 4110–4114. [CrossRef]

50. Vasiljević, J.; Gorjanc, M.; Tomšič, B.; Orel, B.; Jerman, I.; Mozetič, M.; Vesel, A.; Simončič, B. The surface modification of cellulose fibres to create super-hydrophobic, oleophobic and self-cleaning properties. *Cellulose* **2013**, *20*, 277–289. [CrossRef]

51. Ji, H.; Yang, J.; Wu, Z.; Hu, J.; Song, H.; Li, L.; Chen, G. A simple approach to fabricate sticky superhydrophobic polystyrene surfaces. *J. Adhes. Sci. Technol.* **2013**, *27*, 2296–2303. [CrossRef]

52. Vesel, A. Hydrophobization of polymer polystyrene in fluorine plasma. *Mater. Tehnol.* **2011**, *45*, 217–220.

53. Chvátalová, L.; Čermák, R.; Mráček, A.; Grulich, O.; Vesel, A.; Ponížil, P.; Minařík, A.; Cvelbar, U.; Beníček, L.; Sajdl, P. The effect of plasma treatment on structure and properties of poly (1-butene) surface. *Eur. Polym. J.* **2012**, *48*, 866–874. [CrossRef]

54. Grulich, O.; Kregar, Y.; Modic, M.; Vesel, A.; Cvelbar, U.; Mracek, A.; Ponizil, P. Treatment and stability of sodium hyaluronate films in low temperature inductively coupled ammonia plasma. *Plasma Chem. Plasma Process.* **2012**, *32*, 1075–1091. [CrossRef]

55. *ASME B46.1-2009 Surface Texture (Surface Roughness, Waviness, and Lay)*; ASME: New York, NY, USA, 2009.

Article

PVB/ATO Nanocomposites for Glass Coating Applications: Effects of Nanoparticles on the PVB Matrix

Silvia Pizzanelli [1,*], Claudia Forte [1], Simona Bronco [2], Tommaso Guazzini [2], Chiara Serraglini [2] and Lucia Calucci [1]

[1] Istituto di Chimica dei Composti OrganoMetallici, Consiglio Nazionale delle Ricerche-CNR, Via G. Moruzzi 1, 56124 Pisa, Italy; claudia.forte@pi.iccom.cnr.it (C.F.); lucia.calucci@pi.iccom.cnr.it (L.C.)

[2] Istituto per i Processi Chimico-Fisici, Consiglio Nazionale delle Ricerche-CNR, Via G. Moruzzi 1, 56124 Pisa, Italy; simona.bronco@pi.ipcf.cnr.it (S.B.); t.guazza@hotmail.it (T.G.); chiara.serraglini@libero.it (C.S.)

* Correspondence: silvia.pizzanelli@pi.iccom.cnr.it; Tel.: +39-050-315-2549

Received: 28 March 2019; Accepted: 10 April 2019; Published: 12 April 2019

Abstract: Films made of poly(vinyl butyral) (PVB) and antimony-doped tin oxide (ATO) nanoparticles (NPs), both uncoated and surface-modified with an alkoxysilane, were prepared by solution casting at filler volume fractions ranging from 0.08% to 4.5%. The films were characterized by standard techniques including transmission electron microscopy, thermogravimetric analysis and differential scanning calorimetry (DSC). In the polymeric matrix, the primary NPs (diameter ~10 nm) aggregate exhibiting different morphologies depending on the presence of the surface coating. Coated ATO NPs form spherical particles (with a diameter of 300–500 nm), whereas more elongated fractal structures (with a thickness of ~250 nm and length of tens of micrometers) are formed by uncoated NPs. The fraction of the polymer interacting with the NPs is always negligible. In agreement with this finding, DSC data did not reveal any rigid interface and ^{1}H time domain nuclear magnetic resonance (NMR) and fast field-cycling NMR did not show significant differences in polymer dynamics among the different samples. The ultraviolet-visible-near infrared (UV-Vis-NIR) transmittance of the films decreased compared to pure PVB, especially in the NIR range. The solar direct transmittance and the light transmittance were extracted from the spectra according to CEN EN 410/2011 in order to test the performance of our films as plastic layers in laminated glass for glazing.

Keywords: poly(vinyl butyral); antimony doped tin oxide; nanoparticle aggregation; NIR shielding; NMR relaxometry

---

1. Introduction

Polymer nanocomposites are of major scientific and technological interest. In these systems, mechanical properties may be significantly modified at lower loadings compared to microcomposites due to the larger specific surface area [1,2]. Lower loadings facilitate processing and reduce component weight, which makes them industrially attractive. One main goal in the development of high-performance polymer nanocomposites is to obtain a good dispersion of nanoparticles (NPs), which guarantees a high surface area, favoring the interaction with the polymer matrix [3,4]. This is impeded by NP aggregation, which mainly depends on interparticle forces, polymer–NP interactions, and NP shape, as well as on the preparation procedure [5,6].

Antimony-doped tin oxide (ATO) is an optically transparent conducting oxide absorbing in the near infrared (NIR) region as a result of doping, which gives rise to localized surface plasmon resonance. It has been used as filler to increase the electrical conductivity and to provide NIR shielding combined with optical transparency in a variety of polymers, including copolymer lattices containing acrylic units [7–10], poly(acrylate) [11], polyurethane (PU) [12], poly(urethane–acrylate) [13], an epoxy matrix [14], poly(vinyl butyral) (PVB) [15], poly(methyl methacrylate) (PMMA) [16], poly(acrylonitrile) (PAN) [17], and poly(vinyl alcohol) [18]. Studies on the dispersion of ATO NPs in these systems showed that ATO NPs tend to aggregate in network structures when they are bare [7,9,10,13,18]. On the other hand, ATO NPs functionalized with an alkoxy silane were reported to form round sub-micrometric particles in PU [12], PVB [15], and PMMA [16] matrices, and networks of chainlike NPs in PAN [17]. In a study focusing on electrical conductivity [19], ATO/acrylate films were loaded with ATO NPs which were surface-modified using 3-methacryloxypropyltrimethoxysilane (MPS). The ATO NPs gave a fractal type network when a small amount of MPS was used, whereas they formed smaller aggregates in the case of large amounts, indicating that the degree of surface modification affects the morphology of the NP dispersion.

In this work, films of PVB (Figure 1) and ATO NP nanocomposites were prepared by solution casting using NPs, either bare (ATOu) or totally surface modified by MPS (ATOc), at different loading levels. Films were thoroughly characterized using a multi-technique approach, in order to obtain information on the structural and dynamic properties of the different components at the microscopic level, as well as to determine optical properties useful for the application of films as glass coatings. In particular, the dispersion of the NPs in the starting mixtures and in the films was characterized by dynamic light scattering (DLS) and transmission electron microscopy (TEM). The effects of NPs on the thermal properties of the films were investigated by thermogravimetric analysis (TGA) and differential scanning calorimetry (DSC). The fraction of the polymer interacting with the NPs was estimated from TEM measurements. The effects of the filler on the dynamics of the PVB chains were studied by ^{1}H time domain nuclear magnetic resonance (NMR) and fast field-cycling (FFC) NMR relaxometry. These two techniques are less standard in material characterization, although they have often been employed to get insight into structural and dynamic issues in polymer science, including the existence of polymer regions with different mobilities in filled elastomers [20–27]. Finally, considering that the films could be used as plastic layers in the manufacturing of functional safety glass, the effect of both fillers on NIR shielding and optical transparency of the matrix was tested through ultraviolet(UV)-visible(Vis)-NIR transmittance measurements. The influence of the ATO NP surface functionalization on polymer properties was unraveled by comparing data on PVB-ATOu and PVB-ATOc films, while different NP loading levels in the films were exploited to optimize the film composition for applications.

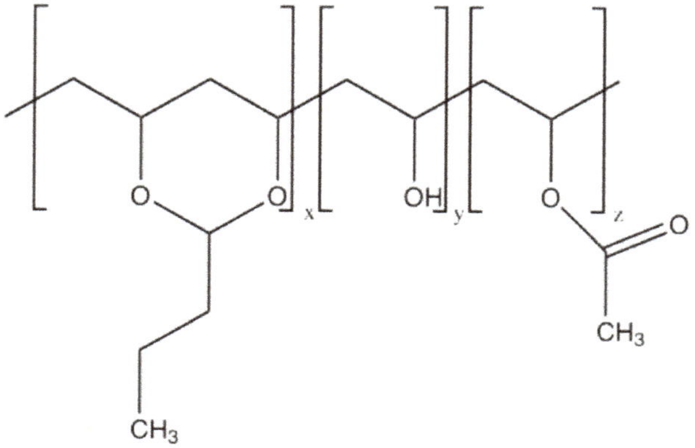

Figure 1. Chemical structure of poly(vinyl butyral) (PVB). For the PVB under investigation, the molar fractions of the vinyl butyral (x), vinyl alcohol (y), and vinyl acetate (z) units are 0.55–0.57, 0.41–0.45, and 0–0.02, respectively.

2. Materials and Methods

2.1. Materials

PVB (trade name Butvar B98®) was purchased from Sigma-Aldrich (St. Louis, MO, USA). The weight average molecular weight, determined using size exclusion chromatography, was 79 kg/mol with a polydispersity of 2.4. The molar fractions of the vinyl butyral, vinyl alcohol, and vinyl acetate units, verified by means of ^{1}H NMR in CDCl$_3$ [28], were 0.55–0.57, 0.41–0.45 and 0–0.02, respectively. Uncoated ATO (ATOu) nanoparticles with a nominal content of Sb$_2$O$_5$ equal to 7–11 wt % were purchased from Sigma-Aldrich. Coated ATO (ATOc) nanoparticles dispersed in ethyl alcohol (10 wt % of ATO) were kindly provided by Kriya Materials (Geleen, The Netherlands) and used as received. The coating agent was MPS.

2.2. Sample Preparation

2.2.1. Suspension of Uncoated ATO Nanoparticles

ATOu nanoparticles were added to ethyl alcohol, ultrasonically dispersed for 30 min, and then subjected to centrifugation (6000 rpm) for 10 min. The dispersed nanoparticles remained suspended in the supernatant, while the undispersed ones precipitated. The amount of nanoparticles in the supernatant (1 wt %) was determined gravimetrically after drying a weighed sample in a rotary evaporator.

2.2.2. Films of PVB and PVB Loaded with ATO Nanoparticles

The films were obtained using the solution casting method, according to a procedure already reported for other PVB composites [29,30]. Briefly, for the PVB blank sample, PVB was dissolved in ethyl alcohol at 75 °C at a concentration of about 4 wt %. Then, the solution was transferred into a Petri Teflon dish and let dry in air for several days. In order to completely remove the solvent, the films were further dried under vacuum (10^{-2} Torr for 10 h). For the loaded samples, PVB was dissolved in ethyl alcohol at 75 °C at a concentration of about 4 wt %. The suspension of ATO nanoparticles, either coated or uncoated, was added to the PVB solution in such an amount that the ATO content in the composite ranged from 0.5 to 23 wt %. The mixture was stirred for 30 min, then transferred into a Petri Teflon dish and let dry in air and then under vacuum. The loaded samples are named PVB-ATOu-i and PVB-ATOc-i, where the letters u and c identify the coated and uncoated oxide, respectively, and i indicates the loading level (0.5, 2, 5, and 23 wt %). The thickness of the obtained films was about 200 µm, as measured by a caliper.

2.3. DLS, TEM, TGA, DSC and UV-Vis-NIR Measurements

DLS was carried out by using a Malvern Zetasizer nano series ZEN1600 (Worcestershire, UK) instrument.

TEM micrographs were collected on suspensions of ATOu and ATOc nanoparticles using a Philips CM12 microscope (Amsterdam, The Netherlands) operating at an accelerating voltage equal to 110 kV and equipped with a Gatan 791 CCD camera. TEM micrographs were also acquired on the films using a Zeiss EM 900 microscope (Oberkochen, Germany) operating at an accelerating voltage equal to 80 kV on ultrathin sections. TEM specimens characterized by a thickness of 40 nm were prepared with a Leica EM FCS cryo-ultracut microtome (Wetzlar, Germany) equipped with a diamond knife.

TGA was performed with a SII TG/DTA 7200 EXSTAR Seiko analyzer (Chiba, Japan), under heating from 30 to 700 °C, at a 10 °C /min rate. Air was fluxed at 200 mL/min during all measurements. The ATOc powder used for the TGA measurement was obtained from the commercial dispersion after evaporation of the ethyl alcohol using a rotary evaporator at 60 °C and 18 mmHg.

DSC experiments were performed on the films using a Seiko SII ExtarDSC7020 calorimeter (Chiba, Japan) with the following thermal protocol: first cooling from 20 to 0 °C; at 0 °C for 2 min; first heating from 0 to 110 °C; 110 °C for 2 min; second cooling from 110 to 0 °C; at 0 °C for 2 min; second heating from 0 to 110 °C; 110 °C for 2 min; third cooling from 110 to 20 °C. The cooling/heating rate was always 10 °C /min except for the last cooling process, when it was fixed to 30 °C /min. The sample amount used for DSC was ~5 mg. Before the DSC measurements, the samples were carefully dried by heating at 100 °C at a pressure of 10^{-2} Torr for 12 h and afterwards kept under a nitrogen atmosphere. The glass transition temperature was determined using the tangents to the measured heat capacities below and above the heat capacity step via the Muse TA Rheo System software (version 3.0). The heat capacity curves of the polymer fraction in the nanocomposites were directly compared using the procedure reported by Cangialosi et al. [31]. Briefly, specific heat capacities of the polymer fraction in the composites, $C_{p,polymer}$, were derived from the measured specific heat capacities, $C_{p,tot}$ (shown in the Supporting Information, Figure S1), by applying the following equation:

$$C_{p,polymer}(T) = \frac{C_{p,tot}(T) - wt\%_{ATO}C_{p,ATO}(T)}{wt\%_{polymer}}, \tag{1}$$

where $C_{p,ATO}$ is the ATO specific heat and $wt\%_{ATO}$ and $wt\%_{polymer}$ are the concentrations of ATO NPs and the polymer, respectively. Next, these specific heat capacities were aligned to the specific heat capacity of the pure PVB above the glass transition by shifting and rotating.

UV-Vis-NIR spectra were recorded with an Agilent Cary 5000 UV-Vis-NIR spectrophotometer (Santa Clara, CA, USA) in the wavelength range between 200 and 3000 nm. In order to evaluate the performances of the samples, light transmittance τ_v (wavelength range 380–780 nm) and solar direct transmittance τ_e (wavelength range 780–2500 nm) were calculated in compliance with CEN EN 410/2011 [32].

2.4. NMR Measurements and Data Analysis

Time domain ^{1}H NMR measurements were performed at 20.7 MHz using a Niumag permanent magnet interfaced with a Stelar PC-NMR console. The temperature of the samples was controlled within ±0.1 °C through a Stelar VTC90 variable temperature controller (Mede, Italy). The ^{1}H 90° pulse duration was 3 μs. On-resonance signals were recorded using the solid echo pulse sequence [33] with an echo time of 14 μs. One hundred and twenty-eight transients were accumulated, and the recycle delay was 3 s. The experiments were performed at 30 and 100 °C, inserting the sample in the pre-heated probe and letting it equilibrate for 10 min. At 100 °C, the signal decay was so slow that field inhomogeneity effects became relevant. In order to exclude these effects, we also performed experiments applying the Carr–Purcell–Meiboom–Gill (CPMG) pulse sequence [34]. The time between successive 180° pulses was 32 μs and the number of transients accumulated was 400.

The ^{1}H longitudinal relaxation times, T_1, were measured at different Larmor frequency values over the 10 kHz–35 MHz range using a Spinmaster FFC-2000 (Stelar srl, Mede, Italy) relaxometer. The measurements were performed using the prepolarized and non-prepolarized pulse sequences below and above 10 MHz, respectively [35,36]. In the former case, a polarizing field of 0.6 T, corresponding to a ^{1}H Larmor frequency of 25.0 MHz, was used. The detection field was 0.5 T, corresponding to a ^{1}H Larmor frequency of 21.5 MHz; the switching time was 3 ms and the probe dead time was 14 μs. The 90° pulse duration was 9.7 μs and 2 scans were accumulated. All the other experimental parameters were optimized for each experiment. Each relaxation trend was acquired with at least 16 values of the variable delay t and was then fitted to the following equation using the SpinMaster fitting procedure.

$$M(t) = M_{relax} + (M_{pol} - M_{relax})exp(-t/T_1), \tag{2}$$

In this equation, M_{pol} and M_{relax} represent the magnetization values in the polarizing and relaxation fields, respectively, with $M_{pol} = 0$ for the non-prepolarized experiments. In all cases, the experimental trends were well reproduced by this equation, with errors on T_1 values lower than 5%. Experiments were performed on heating in the 90–120 °C temperature range, letting the sample temperature equilibrate for 10 min. The temperature of the sample was controlled within ±0.1 °C with a Stelar VTC90 unit.

Before the NMR measurements, the films were heated at 100 °C at a pressure of 10^{-2} Torr for 12 h and afterwards kept under a nitrogen atmosphere. All the measurements were carried out using air as heating gas.

3. Results and Discussion

3.1. Estimate of Coating Degree in ATOc Nanoparticles

Thermogravimetric analysis was performed on the uncoated ATOu and coated ATOc NPs, as well as on the pure MPS coating; results are shown in Figure 2. For both NP samples, the weight loss below 100 °C can be attributed to the desorption of water from the oxide surface, which was more relevant in ATOu, which showed most of the mass loss (2.4%) at temperatures below 100 °C. On the other hand, ATOc exhibited a mass loss lower than 1% below 100 °C, but it lost 9 wt % of its mass between 100 and 700 °C because of the degradation of the coating agent MPS, with the maximum degradation rate occurring at about 330 °C. Since pure MPS shows a TGA profile typical of an evaporating liquid, with the maximum weight loss rate reached at 185 °C (see inset of Figure 2), the behavior of ATOc indicates that MPS is chemically bonded to the surface of ATO particles.

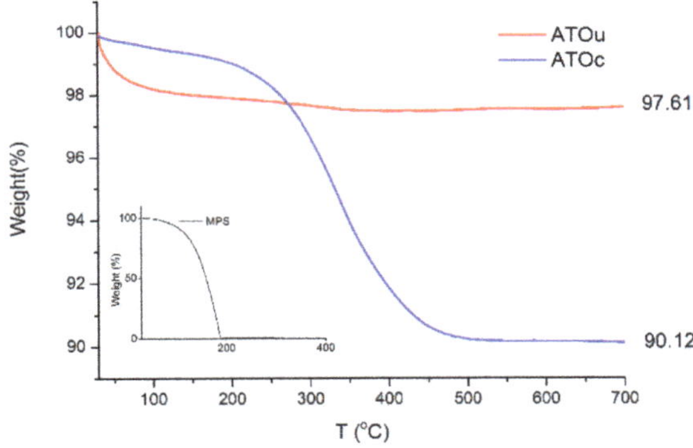

Figure 2. Thermogravimetric analysis (TGA) thermograms of bare antimony-doped tin oxide (ATOu) (red line) and surface modified ATO (ATOc) (blue line). In the inset, the TGA thermogram of pure 3-methacryloxypropyltrimethoxysilane (MPS) is shown for comparison.

The amount of MPS on the ATO surface, Q, could be estimated using the weight loss (wl) in the range between 100 and 700 °C and obtaining the surface area of the remaining 100-wl ATO, S_{ATO}, from the DLS nanoparticle diameter, d, according to the equation

$$S_{ATO} = 6\frac{V_{ATO}}{d} = \frac{6}{d}\frac{100 - wl}{\rho_{ATO}}. \tag{3}$$

Assuming a density for ATO, ρ_{ATO}, of 6.8 g/cm^3, Q ($= wl/(M_{wMPS}{\cdot}S_{ATO})$, where M_{wMPS} is the MPS molecular weight) resulted to be 8 ± 1 µmol/m^2, which is close to the calculated value of 6.9 µmol/m^2 needed for a monolayer when the rod-shaped MPS molecules are perpendicular to the surface [37,38].

3.2. Dispersion of ATO NPs in Ethyl Alcohol and in PVB-ATO Films

Figure 3 shows TEM micrographs of suspensions of ATOu and ATOc NPs in ethyl alcohol. It can be observed that, in both cases, primary particles (with diameters of about 10 nm) tended to form aggregates with average dimensions on the order of a few tens of nanometers; the size distribution was more uniform for ATOc than for ATOu. The analysis of DLS data gave a major population with average diameters of 34 ± 2 nm and 18 ± 2 nm for ATOu and ATOc NPs, respectively, with also a minor population of larger NPs for ATOu. Considering the primary nanoparticle diameter determined from TEM micrographs, we can infer that DLS detected the nanoparticle aggregates present in the dispersions.

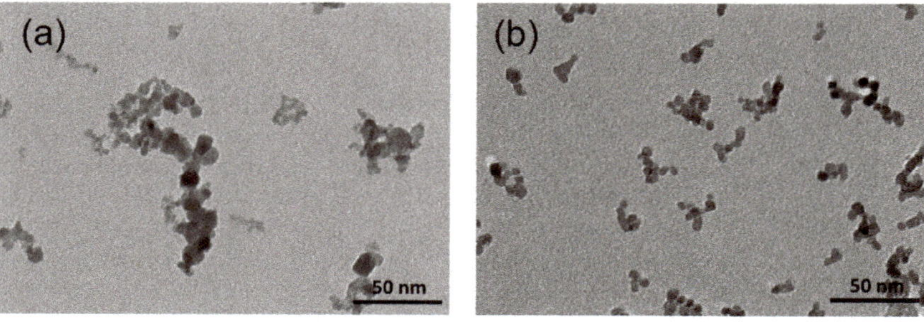

Figure 3. TEM micrographs of suspensions of ATOu (**a**) and ATOc (**b**) in ethyl alcohol.

TEM micrographs of the PVB-ATOu-5 and PVB-ATOc-5 composite films are shown in Figure 4: the images clearly show that the primary nanoparticles form aggregates in the polymer matrix. In the PVB-ATOc-5 sample (Figure 4b,d), round sub-micrometric/micrometric aggregates were observed together with elongated structures characterized by a thickness of up to a few hundreds of nanometers and length of tens of micrometers. The analysis of TEM images indicated a relatively broad distribution of round aggregate sizes (Figure 5) with a mode of 300 nm and a median of 500 nm. It should be noticed that these are upper bounds for the real sizes because superposition effects resulting from the thickness of the TEM specimens may lead to overestimating the real sizes [39]. In the PVB-ATOu-5 sample (Figure 4a,c), essentially only elongated structures were observed with a width of about 250 nm.

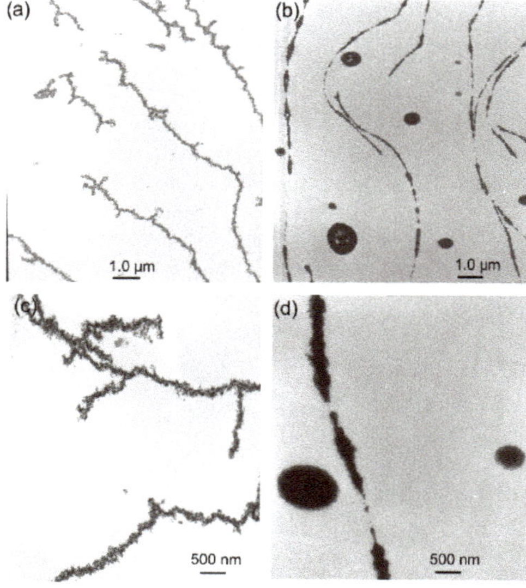

Figure 4. TEM micrographs of the PVB-ATOu-5 (**a,c**) and PVB-ATOc-5 (**b,d**) composite films.

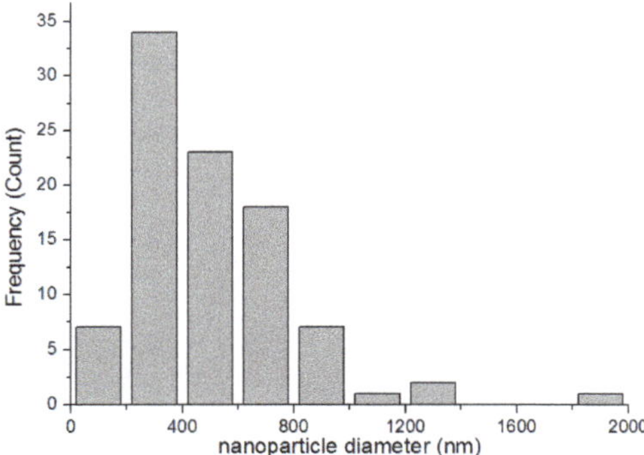

Figure 5. Aggregate size distribution of PVB-ATOc-5.

The morphologies shown by the two samples may be grossly accounted for considering that the films were prepared by casting a dispersion of ATO NPs in an alcoholic solution of PVB. After this dispersion was transferred onto the Teflon dish, all the particles moved about through Brownian motion during ethyl alcohol evaporation, allowing particles to arrange into microstructures in accordance with the evolution of interactions between them during the drying process [9]. In general, in the absence of a repulsive barrier between the NPs, aggregation is expected to be faster and controlled by Brownian motion, thus resulting in tenuous and open fractal structures; on the other hand, if a repulsive barrier exists, aggregation should be slower and controlled by the repulsive interparticle potential, giving rise to denser aggregates which may tend towards round shapes [40]. In our systems, no repulsive barrier exists between the uncoated NPs, whereas a repulsive net interaction between the coated NPs establishes due to the MPS layer [41]. Therefore, our observations reflect the tendency of ATO NPs to aggregate in network structures when they are bare [7,9,10,13,18] and to form round sub-micrometric particles when they are coated [12,15,16].

An estimate of the distance between the surfaces of two neighbouring NP aggregates, h, was obtained from the analysis of TEM data using Equation (4):

$$h = L\left[\left(\frac{\varphi_m}{\varphi}\right)^{\frac{1}{n}} - 1\right], \tag{4}$$

with $n = 3$ or 2 for spherical or cylindrical aggregates, respectively [42]. φ is the aggregate volume fraction, φ_m is the maximum packing fraction (assumed to be 0.64, the value for random close packing of spheres [43] or 0.91 for aligned cylinders packed with hexagonal symmetry [44]), and L is the diameter of the aggregate.

For PVB-ATOu-5, where only chain-like structures were observed, h was estimated setting $n = 2$ and found to be around 2 µm. Assuming the same chain dimensions for PVB-ATOu-23, h was found 700 nm. For PVB-ATOc-5, h values in the range between 1 and 3 µm were estimated, assuming that all aggregates are spherical; since also chain-like structures are present, these represent lower limiting values.

3.3. Thermal Degradation of the PVB-ATO Nanocomposites

Considering that the thermal stability is an important property for nanocomposite applications, the effect of ATO on the thermal degradation of our films under oxidative conditions was also investigated. Thermogravimetric analyses of neat PVB and PVB-ATO composites are shown in Figure 6. The TGA profile of PVB (Figure 6a) is similar to that reported for Butvar B90® in similar conditions [45].

The 2.3% weight loss below 100 °C, shown in the inset, is due to water desorption [46]. This process was shifted to higher temperatures in the composites and was completed by 120 °C. For the main degradation process, the temperature of the maximum degradation rate, T_{max}, obtained from the TG derivative profile, dTG, was 376 °C for PVB. T_{max} shifted to a lower value for the composites, with the effect being more pronounced for samples containing ATOu, as displayed in Table 1. In fact, it has been observed that some oxides [47,48] and acids [49] accelerate the decomposition of PVB.

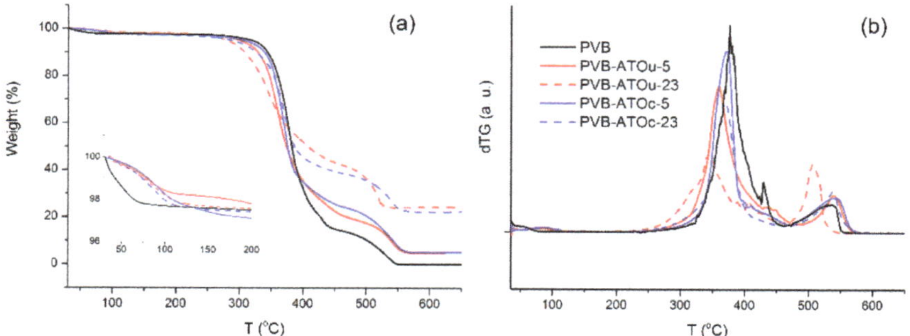

Figure 6. TGA thermograms of PVB (black line), PVB-ATOu-5 (red solid line), PVB-ATOu-23 (red dashed line), PVB-ATOc-5 (blue solid line) and PVB-ATOc-23 (blue dashed line). The remaining weight percentage with respect to the initial value and the dTG profiles are represented in panel (**a**) and (**b**), respectively. The inset in (**a**) shows the weight loss due to water desorption between 30 and 200 °C.

Table 1. ATO nanoparticles (NPs) weight and volume percentages, characteristic degradation temperature T_{max} for the main degradation process and calorimetric T_g values of neat PVB and PVB-ATO composites.

Sample	ATO wt %	ATO vol %	T_{max} (°C)	T_g (°C)
PVB	–	–	376	70.0 ± 0.5
PVB-ATOc-5	5	0.8	371	69.9 ± 0.5
PVB-ATOc-23	23	4.5	370	67.7 ± 0.5
PVB-ATOu-5	5	0.8	360	70.6 ± 0.5
PVB-ATOu-23	23	4.5	343	71.2 ± 0.5

3.4. Interaction between PVB and ATO NPs in the Nanocomposites

3.4.1. Fraction of Interacting Polymer

The fraction of the interacting polymer was estimated from TEM data using Equation (5) [50]:

$$f_{bound\,polymer} = \frac{V_{bound\,polymer}}{V_{total\,polymer}} = \frac{\left(\frac{L}{2}+IL\right)^n - \left(\frac{L}{2}\right)^n}{\left(\frac{L}{2}\right)^n} \cdot \frac{\varphi}{1-\varphi}, \tag{5}$$

where IL is the width of the polymer layer that interacts with the NP aggregates, and the other parameters are defined as in Equation (4). Setting IL = 2 nm, which is a typical reported value [31], and considering spherical aggregates, we found that $f_{bound\,polymer}$ is only 0.2% for PVB-ATOc-23, the nanocomposite with the higher NP loading. If no aggregation occurred, a theoretical value of 8%–10% would be expected. In the case of the PVB-ATOu-23, where the morphology of the aggregates is approximately cylindrical, the fraction of interacting polymer resulted as even lower (0.1%). On the basis of these results, we can state that, as expected, the aggregation of ATO NPs in our systems brings a dramatic reduction of the fraction of interacting polymer $f_{bound\,polymer}$, which indeed becomes negligible.

3.4.2. DSC Data

In order to appreciate the possible effect of the NPs on the thermal properties of the polymer, DSC was conducted on the PVB-ATO composites and on PVB. The DSC thermograms, shown in Figure 7, were aligned according to the procedure outlined in the Materials and Methods section to allow a direct comparison of the heat capacity curves due to the polymer fraction. All the thermograms displayed a baseline shift corresponding to the change in the specific heat from the glass to the melt. For PVB, the glass transition occurred at 70.0 °C. With increasing in the ATO NP content, the glass transition temperature, T_g, slightly increased for the PVB-ATOu composites, whereas it decreased for the PVB-ATOc ones (Table 1). However, since the T_g values changed by only 2–3 °C for a loading level as high as 23 wt %, the observed differences were judged insignificant. As a matter of fact, recent investigations have shown negligible changes in T_g in polymeric nanocomposites, regardless of the polymer–surface interactions, which have been explained by considering that the DSC glass transition is sensitive only to the bulk of the sample and not to the interface [51,52].

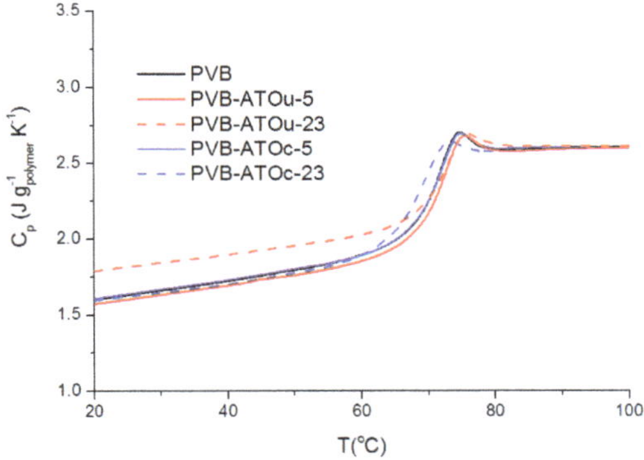

Figure 7. Differential scanning calorimetry (DSC) curves recorded for the PVB and PVB-ATO nanocomposites (second heating cycle). The specific heat capacity of the polymer fraction in the nanocomposites is shown and the curves are aligned to the specific heat capacity of the pure polymer above the glass transition as described in the Materials and Methods section.

In addition, one should notice that for PVB-ATOu-23, the height of the specific heat capacity step at T_g, as determined from the curve in Figure 7, was significantly reduced compared to that observed in the other samples. Considering the low fraction of the interacting polymer estimated for this sample, this reduction is attributed to an enhanced specific heat of the polymer in the glassy state in the nanocomposite compared to the neat PVB, in line with studies reported on other nanocomposites characterized by relatively low loadings [31]. We exclude that the reduction could be attributed to the presence of an immobilized polymer layer close to the NP surface, as done in cases where the specific surface area is considerably larger than that estimated in the present work [53].

3.4.3. NMR Data

The interaction of the polymer with the filler might induce changes in the dynamics of PVB. In order to detect possible differences in the mobility of PVB in the composites with respect to pristine PVB, ^{1}H free induction decays (FIDs) were acquired using the solid echo pulse sequence. This technique allows motions with characteristic frequencies smaller or larger than tens of kHz, corresponding to the static ^{1}H–^{1}H dipolar interaction, to be revealed. In particular, slower motions result in a short decaying signal, whereas faster ones give rise to longer decays. In the glass at 30 °C, no differences among the samples were detected, as shown in Figure 8a. In all cases, the decay occurred within 30 µs, indicating that the hydrogen atoms of the polymer were in a rigid glassy environment with characteristic motional frequencies smaller than tens of kHz. At 100 °C, the decays occurred over a much longer time scale for all samples, with the signals persisting after 200 μs from the echo maximum (Figure 8b). At this temperature, CPMG experiments, in which effects on the signal decay due to field inhomogeneities are avoided, confirmed the similar behavior for all the samples (data not shown). These findings indicate a softening of the polymer above T_g, with no detectable residual glassy domains at the interface between the NPs and the polymer, in agreement with the negligible interacting polymer fraction estimated by TEM. The very similar results obtained for pristine PVB and PVB-ATO at different loadings also show that the polymer dynamics was not appreciably affected by the filler.

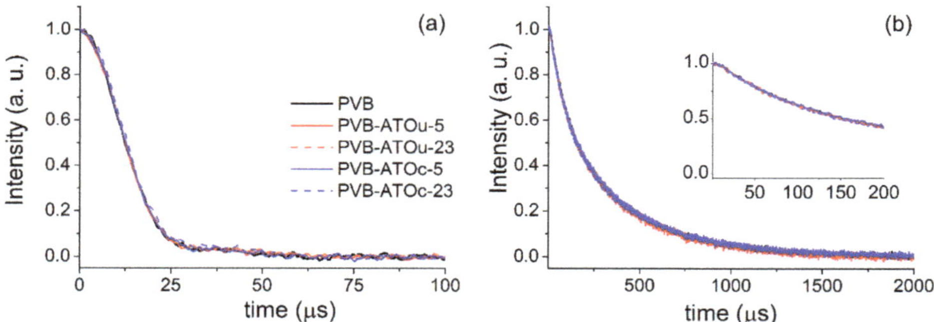

Figure 8. Comparison of the free induction decays (FIDs) exhibited by PVB and by the indicated PVB-ATO composites at 30 (**a**) and 100 °C (**b**). The FIDs were recorded by applying the solid echo pulse sequence. t indicates the time elapsed from the echo maximum. The signal intensities were scaled to the same value at $t = 0$ in order to facilitate the comparison of the decay rates.

It is worthy of note that similar signal decays might be observed notwithstanding the different frequency distribution of chain motions. In order to verify possible effects on the motional frequency distribution, a more detailed analysis of the motional behavior was performed by exploiting ^{1}H FFC NMR relaxometry. Indeed, this technique allows polymer dynamics to be investigated by measuring the dependence of the proton longitudinal relaxation rates $1/T_1$ on the Larmor frequency; i.e., the so-called NMR relaxation dispersion (NMRD) curve. This technique is sensitive to motions characterized by frequencies ranging from 10 kHz to tens of MHz. In a previous study on pure PVB [54], FFC NMR relaxometry allowed different motional processes to be identified above T_g: segmental dynamics with characteristic frequencies of 10^5–10^6 Hz and faster local motions with frequencies higher than 10^7 Hz. In the present work, the NMRD curves were acquired on the PVB-ATO composites in the temperature range between 90 and 120 °C; representative curves are shown in Figure 9. The acquired NMRD curves clearly showed that the motional processes occurring in PVB between 10^5 and 10^7 Hz range were not affected by the presence of ATO nanoparticles.

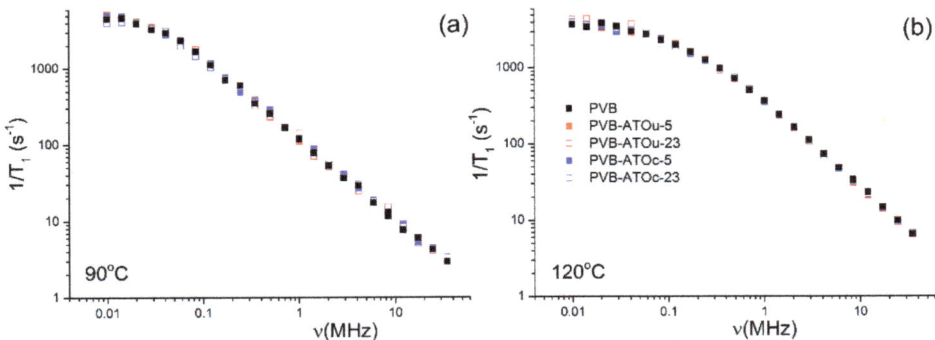

Figure 9. Comparison of the nuclear magnetic resonance relaxation dispersion (NMRD) curves exhibited by PVB and by the composites at (**a**) 90 °C and (**b**) 120 °C.

3.5. Optical Properties of PVB-ATO Composites

3.5.1. UV-Vis-NIR Transmittance

To investigate how the content of ATO NPs affects the optical properties of the composites, we measured the UV-Vis-NIR transmittance of PVB loaded with both ATOu and ATOc NPs at filler fractions ranging from 0.5 to 5 wt %; the spectra are shown in Figure 10. The high shielding efficiency in the NIR region is due to ATO absorption [55] and has been previously observed for other nanocomposites containing ATO [12,14]. The transmittance in the visible region decreased with increased the ATO content and, at the same loading level, it was more pronounced for the composites containing ATOc compared to ATOu. This can be attributed to the fact that the morphology and dimensions of the ATOc NP aggregates favor light scattering. In fact, the intensity of the scattered light is known to increase drastically when the size of the particles is on the order of the wavelength of visible light [56]. In PVB-ATOc-5, the aggregate sizes were within the visible light wavelength range for more than half of the aggregates (Figure 5), whereas the width of the elongated NP aggregates observed in PVB-ATOu-5 (Figure 4c) was only about 250 nm.

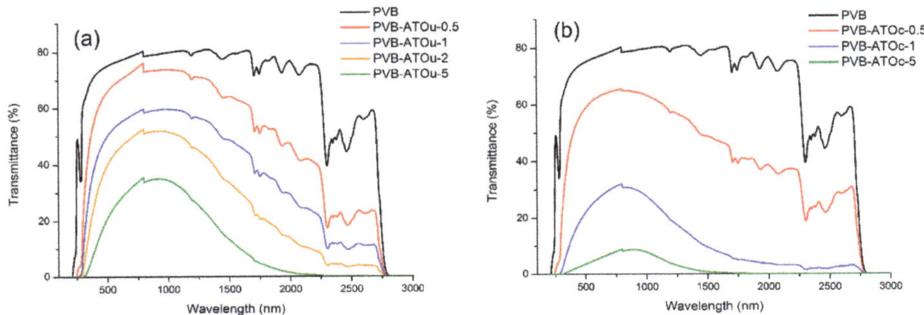

Figure 10. UV-Vis-NIR spectra of PVB-ATOu (**a**) and of PVB-ATOc (**b**) at different loadings. The spectrum of neat PVB is reported for comparison. The abrupt jump at 800 nm is due to lamp change.

In order to quantitatively compare the behavior of the films towards visible and near infrared light, the transmittance values of each sample were averaged over each wavelength range and are reported in Table 2. Considering that glasses in car windshields are required to have a visible transmittance larger than 70% [57], we can state that only PVB-ATOu-0.5 resulted to be almost optically transparent.

In addition, this sample showed a 25% decrease in NIR transmittance with respect to neat PVB. These findings make PVB-ATOu-0.5 a promising material for the industry of functional safety glass.

Table 2. Transmittance values of PVB and PVB-ATO films averaged over the wavelength range of visible and near infrared light.

Wt % Loading	Transmittance (%)			
	PVB-ATOu		PVB-ATOc	
	NIR (780–2400 nm)	Vis (380–780 nm)	NIR (780–2400 nm)	Vis (380–780 nm)
0	76	75	76	75
0.5	57	68	47	60
2	30	43	12	25
5	14	25	2	5

3.5.2. Solar and Luminous Characteristics

The performance of the investigated films as plastic layers in low emissive laminated glasses for glazing in energy-saving buildings was evaluated on the basis of visible light transmittance τ_v and solar direct transmittance τ_e, as defined in CEN EN 410/2011 [32]. τ_v represents the glazing system capacity to diffuse the natural light indoors as perceived by the human eye. A higher τ_v value ensures a better capacity, which is desirable for saving electric energy in the daytime. τ_e contributes to the total solar energy transmittance (solar factor) together with a heat transfer factor. In this case, a lower solar factor is desirable because it means a lower heat penetration inside and a consequent reduction in air-conditioning operating costs. τ_v and τ_e values for the PVB-ATO films at different loading levels were extracted from the UV-Vis-NIR spectral data and are shown in Figure 11. The trends of τ_v and τ_e are similar to those reported in Table 2 for Vis and NIR transmittances, respectively, and confirm that PVB-ATOu-0.5 is the best-performing material. In fact, it induces a significant reduction of τ_e compared to pure PVB while leaving τ_v at acceptably high values.

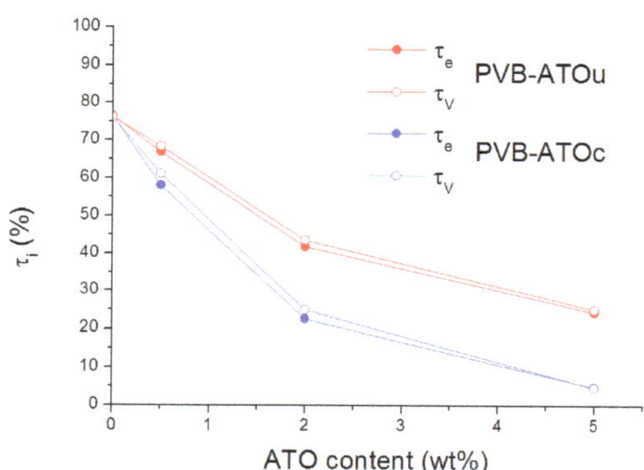

Figure 11. Visible light transmittance, τ_v, and solar direct transmittance, τ_e, as a function of the ATO content.

4. Conclusions

Films of PVB-ATO composites prepared by solution casting were studied using a multi-technique approach. The primary ATO NPs were found to aggregate in different morphologies depending on the presence of a coating on the NP surface: elongated fractal structures characterized by a thickness of about 250 nanometers and a length of tens of micrometers and spherical aggregates with sizes of hundreds of nm were found for the uncoated and coated NPs, respectively. This aggregation, reducing the filler surface area, resulted in a negligible fraction of polymer interacting with the NPs, as also suggested by the lack of significant effects on the polymer glass transition temperature and on the polymer dynamics, as determined by calorimetric and NMR relaxometry experiments. On the other hand, ATO NPs significantly modified the optical properties of the films, with their presence resulting in a high shielding efficiency in the NIR region together with a reduced transmittance in the visible region. These effects were more pronounced for the composites containing ATOc and were ascribed to the comparable size of the NP aggregates with the light wavelength. PVB-ATOu-0.5 showed the best performance in terms of light transmittance and solar direct transmittance, which are both relevant parameters for glazing in building.

The applied thorough multi-technique approach was extremely useful in unraveling the effect of ATO NPs on PVB properties which are fundamental for potential industrial applications, such as glass transition temperature, thermal degradation and response to UV, Vis and NIR light, and allowed these properties to be correlated to structural and dynamic properties of the different composite components at the microscopic level.

Supplementary Materials: The following are available online at http://www.mdpi.com/2079-6412/9/4/247/s1, Figure S1: Specific heat capacity of nanocomposites in respect to sample mass. Specific heat capacities of ATOu and ATOc are also shown.

Author Contributions: Conceptualization, S.P. and S.B.; Methodology, T.G., S.B., S.P., L.C.; Software, T.G., C.S. and S.P.; Validation, C.S., T.G., S.P. and L.C.; Formal Analysis, T.G., C.S. and S.P.; Investigation, T.G., C.S., S.P. and L.C.; Resources, C.F. and S.B.; Data Curation, S.P. and S.B.; Writing—Original Draft Preparation, S.P.; Writing—Review & Editing, S.P., C.F., L.C., T.G. and S.B.; Visualization, S.P., T.G. and C.S.; Supervision, S.P., L.C. and S.B.; Project Administration, S.P. and S.B.; Funding Acquisition, S.B.

Funding: This work was partially supported by Regione Toscana in the framework of the project SELFIE (Bando FAR-FAS 2014-Programma PAR FAS 2007-2013- Linea d'Azione 1.1).

Acknowledgments: We acknowledge Marco Geppi for kindly letting us use the 20.7 MHz spectrometer and Maria Cristina Righetti for helpful discussion. In addition, we would like to acknowledge the contribution of the COST Action CA15209 (Eurelax: European Network on NMR Relaxometry).

Conflicts of Interest: The authors declare no conflict of interest.

References

1. Alexandre, M.; Dubois, P. Polymer-layered silicate nanocomposites: Preparation, properties and uses of a new class of materials. *Mater. Sci. Eng. R Rep.* **2000**, *28*, 1–63. [CrossRef]
2. Winey, K.I.; Vaia, R.A. Polymer nanocomposites. *MRS Bull.* **2007**, *32*, 314–322. [CrossRef]
3. Krishnamoorti, R. Strategies for dispersing nanoparticles in polymers. *MRS Bull.* **2007**, *32*, 341–347. [CrossRef]
4. Mackay, M.E.; Tuteja, A.; Duxbury, P.M.; Hawker, C.J.; Van Horn, B.; Guan, Z.; Chen, G.; Krishnan, R.S. General strategies for nanoparticle dispersion. *Science* **2006**, *311*, 1740–1743. [CrossRef] [PubMed]
5. Min, Y.; Akbulut, M.; Kristiansen, K.; Golan, Y.; Israelachvili, J. The role of interparticle and external forces in nanoparticle assembly. *Nat. Mater.* **2008**, *7*, 527–538. [CrossRef]
6. Natarajan, B.; Li, Y.; Deng, H.; Brinson, L.C.; Schadler, L.S. Effect of interfacial energetics on dispersion and glass transition temperature in polymer nanocomposites. *Macromolecules* **2013**, *46*, 2833–2841. [CrossRef]
7. Sun, J.; Gerberich, W.W.; Francis, L.F. Electrical and optical properties of ceramic-polymer nanocomposite coatings. *J. Polym. Sci. Part B: Polym. Phys.* **2003**, *41*, 1744–1761. [CrossRef]
8. Sun, J.; Velamakanni, B.V.; Gerberich, W.W.; Francis, L.F. Aqueous Latex/Ceramic Nanoparticle dispersions: Colloidal stability and coating properties. *J. Colloid Interface Sci.* **2004**, *280*, 387–399. [CrossRef] [PubMed]
9. Wang, Y.; Anderson, C. Formation of thin transparent conductive composite films from aqueous colloidal dispersions. *Macromolecules* **1999**, *32*, 6172–6179. [CrossRef]
10. Wakabayashi, A.; Sasakawa, Y.; Dobashi, T.; Yamamoto, T. Optically transparent conductive network formation induced by solvent evaporation from tin-oxide-nanoparticle suspensions. *Langmuir* **2007**, *23*, 7990–7994. [CrossRef]

11. Kleinjan, W.E.; Brokken-Zijp, J.C.M.; van de Belt, R.; Chen, Z.; de With, G. Antimony-doped tin oxide nanoparticles for conductive polymer nanocomposites. *J. Mater. Res.* **2008**, *23*, 869–880. [CrossRef]

12. Zhou, H.; Wang, H.; Tian, X.; Zheng, K.; Cheng, Q. Preparation and properties of waterborne polyurethane/antimony doped Tin oxide nanocomposite coatings via Sol–Gel reactions. *Polym. Compos.* **2014**, *35*, 1169–1175. [CrossRef]

13. Wu, K.; Xiang, S.; Zhi, W.; Bian, R.; Wang, C.; Cai, D. Preparation and characterization of UV curable waterborne Poly(Urethane-Acrylate)/Antimony doped Tin Oxide thermal insulation coatings by Sol-Gel process. *Prog. Org. Coat.* **2017**, *113*, 39–46. [CrossRef]

14. Mei, S.-G.; Ma, W.-J.; Zhang, G.-L.; Wang, J.-L.; Yang, J.-H.; Li, Y.-Q. Transparent ATO/Epoxy nanocomposite coating with excellent thermal insulation property. *Micro Nano Lett.* **2012**, *7*, 12–14. [CrossRef]

15. Zhang, G.; Yan, W.; Jiang, T. Fabrication and thermal insulating properties of ATO/PVB nanocomposites for energy saving glass. *J. Wuhan Univ. Technol. Mater. Sci. Ed.* **2013**, *28*, 912–915. [CrossRef]

16. Pan, W.; Zhang, H.; Chen, Y. Electrical and mechanical properties of PMMA/Nano-ATO composites. *J. Mater. Sci. Tech. (Shenyang, China)* **2009**, *25*, 247–250.

17. Pan, W.; Zou, H. Characterization of PAN/ATO nanocomposites prepared by solution blending. *Bull. Mater. Sci.* **2008**, *31*, 807–811. [CrossRef]

18. Pan, W.; He, X.; Chen, Y. Preparation and characterization of Poly (Vinyl Alcohol)/Antimony-Doped Tin Oxide nanocomposites. *Int. J. Polym. Mater. Polym. Biomater.* **2011**, *60*, 223–232. [CrossRef]

19. Posthumus, W.; Laven, J.; de With, G.; van der Linde, R. Control of the electrical conductivity of composites of antimony Doped Tin Oxide (ATO) nanoparticles and acrylate by grafting of 3-Methacryloxypropyltrimethoxysilane (MPS). *J. Colloid Interface Sci.* **2006**, *304*, 394–401. [CrossRef] [PubMed]

20. Kaufman, S.; Slichter, W.P.; Davis, D. Nuclear magnetic resonance study of rubber–carbon black interactions. *J. Polym. Sci. A-2, Polym. Phys.* **1971**, *9*, 829–839. [CrossRef]

21. Litvinov, V.; Steeman, P. EPDM−Carbon Black Interactions and the reinforcement mechanisms, As studied by Low-Resolution ^{1}H NMR. *Macromolecules* **1999**, *32*, 8476–8490. [CrossRef]

22. Dreiss, C.A.; Cosgrove, T.; Benton, N.J.; Kilburn, D.; Alam, M.A.; Schmidt, R.G.; Gordon, G.V. Effect of crosslinking on the mobility of PDMS filled with polysilicate nanoparticles: positron Lifetime, rheology and NMR relaxation studies. *Polymer (Guildf)* **2007**, *48*, 4419–4428. [CrossRef]

23. Papon, A.; Montes, H.; Hanafi, M.; Lequeux, F.; Guy, L.; Saalwächter, K. Glass-Transition temperature gradient in nanocomposites: evidence from nuclear magnetic resonance and differential scanning calorimetry. *Phys. Rev. Lett.* **2012**, *108*, 1–5. [CrossRef] [PubMed]

24. Kim, S.Y.; Meyer, H.W.; Saalwächter, K.; Zukoski, C.F. Polymer dynamics in PEG-Silica nanocomposites: Effects of polymer molecular weight, temperature and solvent dilution. *Macromolecules* **2012**, *45*, 4225–4237. [CrossRef]

25. Valle Iulianelli, G.C.; dos S. David, G.; dos Santos, T.N.; Sebastião, P.J.O.; Tavares, M.I.B. Influence of TiO2 nanoparticle on the thermal, morphological and molecular characteristics of PHB matrix. *Polym. Test.* **2018**, *65*, 156–162. [CrossRef]

26. Brito, L.M.; Chávez, F.V.; Bruno Tavares, M.I.; Sebastião, P.J.O. Molecular dynamic evaluation of starch-PLA blends nanocomposite with organoclay by proton NMR relaxometry. *Polym. Test.* **2013**, *32*, 1181–1185. [CrossRef]

27. Krzaczkowska, J.; Strankowski, M.; Jurga, S.; Jurga, K.; Pietraszko, A. NMR dispersion studies of Poly(Ethylene Oxide)/Sodium montmorillonite nanocomposites. *J. Non-Cryst. Solids* **2010**, *356*, 945–951. [CrossRef]

28. Fernández, M.D.; Fernández, M.J.; Hoces, P. Synthesis of Poly (Vinyl Butyral)s in homogeneous phase and their thermal properties. *J. Appl. Polym. Sci.* **2006**, *102*, 5007–5017. [CrossRef]

29. Gupta, S.; Seethamraju, S.; Ramamurthy, P.C.; Madras, G. Polyvinylbutyral based hybrid organic/inorganic films as a moisture barrier material. *Ind. Eng. Chem. Res.* **2013**, *52*, 4383–4394. [CrossRef]

30. Roy, A.S.; Gupta, S.; Seethamraju, S.; Madras, G.; Ramamurthy, P.C. Impedance spectroscopy of novel hybrid composite films of Polyvinylbutyral (PVB)/Functionalized mesoporous silica. *Compos. Part B Eng.* **2014**, *58*, 134–139. [CrossRef]

31. Cangialosi, D.; Boucher, V.M.; Alegría, A.; Colmenero, J. Enhanced physical aging of polymer nanocomposites: The key role of the area to volume ratio. *Polymer (Guildf)* **2012**, *53*, 1362–1372. [CrossRef]

32. IS EN 410 Glass in Building. *Determination of Luminous and Solar Characteristics of Glazing*; National Standards Authority of Ireland: Dublin, Ireland, 2011.

33. Powles, J.G.; Strange, J.H. Zero Time resolution nuclear magnetic resonance transient in solids. *Proc. Phys. Soc. Lond.* **1963**, *82*, 6–15. [CrossRef]

34. Meiboom, S.; Gill, D. Modified spin—Echo method for measuring nuclear relaxation times. *Rev. Sci. Instrum.* **1958**, *29*, 688–691. [CrossRef]

35. Anoardo, E.; Galli, G.; Ferrante, G. Fast-Field-Cycling NMR: Applications and instrumentation. *Appl. Magn. Reson.* **2001**, *20*, 365–404. [CrossRef]

36. Kimmich, R.; Anoardo, E. Field-Cycling NMR Relaxometry. *Prog. Nucl. Magn. Reson. Spectrosc.* **2004**, *44*, 257–320. [CrossRef]

37. Miller, J.D.; Ishida, H. Quantitative Monomolecular coverage of inorganic particulates by methacryl-functional silanes. *Surf. Sci.* **1984**, *148*, 601–622. [CrossRef]

38. Posthumus, W.; Magusin, P.C.M.M.; Brokken-Zijp, J.C.M.; Tinnemans, A.H.A.; van der Linde, R. Surface modification of oxidic nanoparticles using 3-methacryloxypropyltrimethoxysilane. *J. Colloid Interface Sci.* **2004**, *269*, 109–116. [CrossRef]

39. Hajji, P.; David, L.; Gerard, J.F.; Pascault, J.P.; Vigier, G. Synthesis, structure, and morphology of polymer–silica hybrid nanocomposites based on hydroxyethyl Methacrylate. *J. Polym. Sci. Part B Polym. Phys.* **1999**, *37*, 3172–3187. [CrossRef]

40. Meakin, P.; Jullien, R. The effects of restructuring on the geometry of clusters formed by diffusion-limited, ballistic, and reaction-limited cluster-cluster aggregation. *J. Chem. Phys.* **1988**, *89*, 246–250. [CrossRef]

41. Miller, K.T.; Zukoski, C.F. The mechanics of nanoscale suspensions. In *Semiconductor Nanoclusters—Physical, Chemical, and Catalytic Aspects*; Kamat, P.V., Meisel, D., Eds.; Elsevier: Amsterdam, The Netherlands, 1997; Volume 103, pp. 23–55.

42. Hao, T.; Riman, R.E. Calculation of interparticle spacing in colloidal systems. *J. Colloid Interface Sci.* **2006**, *297*, 374–377. [CrossRef]

43. Bernal, J.D.; Mason, J. Packing of spheres: Co-ordination of randomly packed spheres. *Nature* **1960**, *188*, 910–911. [CrossRef]

44. Bezdek, A.; Kuperberg, W. Maximum density space packing with congruent circular cylinders of infinite length. *Mathematika* **1990**, *37*, 74–80. [CrossRef]

45. Liau, L.C.-K.; Chien, Y.-C. Kinetic investigation of ZrO_2, Y_2O_3, and Ni on Poly (Vinyl Butyral) thermal degradation using nonlinear heating functions. *J. Appl. Polym. Sci.* **2006**, *102*, 2552–2559. [CrossRef]

46. Carini, G., Jr.; Bartolotta, A.; Carini, G.; D'Angelo, G.; Federico, M.; Di Marco, G. Water-Driven segmental cooperativity in polyvinyl butyral. *Eur. Polym. J.* **2018**, *98*, 172–176. [CrossRef]

47. Masia, S.; Calvert, P.D.; Rhine, W.E.; Bowen, H.K. Effect of oxides on binder burnout during ceramics processing. *J. Mater. Sci.* **1989**, *24*, 1907–1912. [CrossRef]

48. Saravanan, S.; Gupta, S.; Ramamurthy, P.C.; Madras, G. Effect of silane functionalized alumina on Poly(Vinyl Butyral) nanocomposite films: Thermal, mechanical, and moisture barrier studies. *Polym. Compos.* **2014**, *35*, 1426–1435. [CrossRef]

49. Liu, R.; He, B.; Chen, X. Degradation of Poly(Vinyl Butyral) and its stabilization by bases. *Polym. Degrad. Stab.* **2008**, *93*, 846–853. [CrossRef]

50. Becker, C.; Krug, H. Schmidt, tailoring of thermomechanical properties of thermoplastic nanocomposites by surface modification of nanoscale silica particles. *Mater. Res. Soc. Symp. Proc.* **1996**, *435*, 237–242. [CrossRef]

51. Moll, J.; Kumar, S.K. Glass transitions in highly attractive highly filled polymer nanocomposites. *Macromolecules* **2012**, *45*, 1131–1135. [CrossRef]

52. Holt, A.P.; Griffin, P.J.; Bocharova, V.; Agapov, A.L.; Imel, A.E.; Dadmun, M.D.; Sangoro, J.R.; Sokolov, A.P. Dynamics at the Polymer/Nanoparticle Interface in Poly(2-vinylpyridine)/Silica Nanocomposites. *Macromolecules* **2014**, *47*, 1837–1843. [CrossRef]

53. Sargsyan, A.; Tonoyan, A.; Davtyan, S.; Schick, C. The amount of immobilized polymer in PMMA SiO_2 nanocomposites determined from calorimetric data. *Eur. Polym. J.* **2007**, *43*, 3113–3127. [CrossRef]

54. Pizzanelli, S.; Prevosto, D.; Guazzini, T.; Bronco, S.; Forte, C.; Calucci, L. Dynamics of Poly (Vinyl Butyral) studied using dielectric spectroscopy and ^{1}H NMR relaxometry. *Phys. Chem. Chem. Phys.* **2017**, *19*, 31804–31812. [CrossRef]

55. Lee, H.Y.; Cai, Y.; Bi, S.; Liang, Y.N.; Song, Y.; Hu, X.M. A Dual-Responsive nanocomposite toward climate-adaptable solar modulation for energy-saving smart windows. *ACS Appl. Mater. Interfaces* **2017**, *9*, 6054–6063. [CrossRef]

56. Kerker, M. *The Scattering of Light and Other Electromagnetic Radiation*, 1st ed.; Academic Press: New York, NY, USA, 1969.
57. ISO 3917. *2016 Road Vehicles-Safety Glazing Materials-Test Methods for Resistance to Radiation, High Temperature, Humidity, Fire and Simulated Weathering*; International Organization for Standardization: Geneva, Switzerland, 2016.

 *coatings*

Article

Tribological Performance of PVD Film Systems Against Plastic Counterparts for Adhesion-Reducing Application in Injection Molds

Wolfgang Tillmann, Nelson Filipe Lopes Dias *, Dominic Stangier and Nikolai Gelinski

Institute of Materials Engineering, TU Dortmund University, Leonhard-Euler-Str. 2, D-44227 Dortmund, Germany; wolfgang.tillmann@tu-dortmund.de (W.T.); dominic.stangier@tu-dortmund.de (D.S.); nikolai.gelinski@tu-dortmund.de (N.G.)

* Correspondence: filipe.dias@tu-dortmund.de; Tel.: +49-231-755-5139

Received: 2 September 2019; Accepted: 16 September 2019; Published: 17 September 2019

Abstract: The deposition of physical vapor deposition (PVD) hard films is a promising approach to enhance the tribological properties of injection molds in plastic processing. However, the adhesion is influenced by the pairing of PVD film and processed plastic. For this reason, the friction behavior of different PVD films against polyamide, polypropylene, and polystyrene was investigated in tribometer tests by correlating the relation between the roughness and the adhesion. It was shown that the dispersive and polar surface energy have an impact on the work of adhesion. In particular, Cr-based nitrides with a low polar component exhibit the lowest values ranging from 65.5 to 69.4 mN/m when paired with the polar polyamide. An increased roughness leads to a lower friction due to a reduction of the adhesive friction component, whereas a higher work of adhesion results in higher friction for polyamide and polypropylene. Within this context, most Cr-based nitrides exhibited coefficients of friction below 0.4. In contrast, polystyrene leads to a friction-reducing material transfer. Therefore, a customized deposition of the injection molds with an appropriated PVD film system should be carried out according to the processed plastic.

Keywords: physical vapor deposition; CrN; CrAlN; CrAlSiN; Al_2O_3; amorphous carbon; tribological properties; polyamide; polypropylene; polystyrene

1. Introduction

Injection molding is a cyclic process for the cost-effective mass production of plastic components with complex geometries at relatively low processing temperatures [1]. The cycle sequence consists of mold closing, injection of melt, packing to compensate the shrinkage, cooling, and mold opening when ejecting the parts [2]. Depending on the processed plastic, the injection molding tools are exposed to high tribological loads and subjected to corrosion and pitting [3]. Moreover, the molded parts tend to stick on the surface of the tool, resulting in a buildup and restricting the productivity of the process as well as the quality of the plastic parts [4]. Within this context, lower adhesion strengths between injection mold and molded part also enable to reduce the ejection forces, which allows reduction of the dimensions of the ejection system within the mold, thus exploiting the attained space by expanding the cooling circuit [5]. As a consequence, the cooling time is reduced and higher production rates can be achieved [6]. Therefore, a suitable surface modification of the injection molding tools is necessary to optimize the tribological properties [7].

A promising approach is the deposition of PVD (physical vapor deposition) films with low adhesion properties to plastic parts [8], a high wear resistance [9,10], and a high corrosion resistance [11]. Within this context, several research works deposited different PVD film systems and investigated either the surface free energy and adhesion behavior in abstracted laboratory tests [12,13] or determined the ejection forces in application-oriented tests [14,15]. Bagcivan et al. conducted high-temperature contact angle measurements on chromium aluminum oxynitride (CrAlON) films and uncoated American Iron and Steel Institute (AISI) 420 with polymethyl methacrylate and reported larger contact angles for the oxynitride coatings [12]. In addition, lower coefficients of friction for CrAlON were observed when compared to AISI 420 and it was concluded that the adhesion behavior has an impact on the friction [12]. Bobzin et al. also investigated the wetting behavior of uncoated AISI 420 and chromium aluminum nitride (CrAlN) as well as CrAlON films with polycarbonate (PC) at 280 °C and observed the largest contact angles for the oxynitridic films [13]. Furthermore, an adhesion test in compliance with DIN EN ISO 4624 was performed to determine the adhesion tensile strength of solidified PC on the surface systems [14]. Lower adhesion tensile strengths for the films were reported in comparison to uncoated steel, but no significant difference between CrAlON and CrAlN was detected. Sasaki et al. deposited titanium nitride (TiN), chromium nitride (CrN), diamond-like carbon (DLC) as well as tungsten carbon carbide (WC/C) and determined the ejection force when removing molded polypropylene (PP) and polyethylene terephthalate (PET) parts from the core [15]. Lower ejection forces for the PVD films were noted in comparison to the uncoated mold, but an explanation for this behavior was not given. Burkard et al. investigated the correlation between film system and plastic counterpart on the ejection force by testing CrN, WC/C, TiN, titanium aluminum nitride (TiAlN) with polyamide (PA), PC, acrylonitrile butadiene styrene as well as polyoxymethylene and observed a different behavior for each plastic counterpart [16]. Burkard et al. concluded that it is not possible to make a general statement, concerning which film system is suitable to reduce the ejection force as this depends on its pairing with the plastic part.

For this reason, the pairing of the PVD film system and the processed plastic part has a key role in the tribological performance of the PVD hard films. However, it should be noted that besides the chemical composition of the PVD films, the roughness profile also has an influence on the friction component and should therefore be taken into consideration as well [17]. Thus, a total amount of eight PVD film systems with different chemical composition and mechanical properties were analyzed regarding their tribological behavior against PA, PP, and polystyrene (PS). The surface energy of the PVD films was determined in laboratory tests and the work of adhesion with PA, PP, and PS was calculated theoretically. The friction against the plastic counterparts was analyzed in tribometer tests. Based on the results, a correlation between the roughness and the adhesion properties was identified, which has an effect on the resulting friction behavior. A comprehensive understanding of this relation allows choosing a suitable PVD film to reduce the adhesive friction component in injection molding.

2. Materials and Methods

Quenched and tempered AISI H11 steel with a hardness of 7.0 ± 0.3 GPa was used as substrate material and polished prior to the depositions. The metallographical preparation consisted of grinding with SiC papers from 320 to 1200 grits for 3 min each and polishing using diamond suspension from 9 to 1 μm gran size for 6 min each. The steel is characterized by a high toughness, high-temperature strength, and high thermal shock resistance and is utilized as a tool steel for injection molds in plastics processing [18]. The deposition of the PVD films was carried out by means of an industrial magnetron sputtering device CC800/9 Custom (CemeCon AG, Würselen, Germany). A total of eight different film systems from the groups of Cr-based nitrides, oxides, and DLC films were selected for the experiments (see Figure 1). Since the adhesion behavior depends on the pairing between the PVD film and the molded plastic part, an appropriated PVD film system for adhesion-reducing application in injection molding of PA, PP, and PS needs to be determined first. For this purpose, PVD films of industrial relevance were selected to cover a broad range of systems with different chemical composition in order to obtain distinct surface energies and consequently a different adhesion behavior with PA, PP, and PS. Moreover, these films were deposited either in direct current magnetron sputtering (dcMS), mid-frequency magnetron sputtering (mfMS), or high-power impulse magnetron sputtering (HiPIMS) mode. Within this context, CrN and CrAlN films were synthesized by dcMS as well as

HiPIMS. This allows synthesis of film systems with a similar chemical composition, but with different topographical structures caused by the sputtering technique. This allows consideration of the influence of the surface structure on the tribological behavior of the films against the counterpart. Furthermore, chromium aluminum silicon nitride (CrAlSiN) was deposited in dcMS mode. From the group of the oxidic films, amorphous aluminum oxide (Al_2O_3) was reactively sputtered from aluminum targets in the mfMS mode. All film systems were applied as a monolayer on the steel substrate. Besides the monolayer systems, hydrogen-free amorphous carbon (a-C) and hydrogenated amorphous carbon (a-C:H) films were deposited by mfMS. In order to improve the adhesion of the carbon layers to the substrate, a graded chemical transition with CrN, chromium carbonitride (CrCN), and chromium carbide (CrC) interlayers were sputtered prior to the a-C and a-C:H top layer. The graded systems as well as the deposition process are described in more detail by Hoffmann [19].

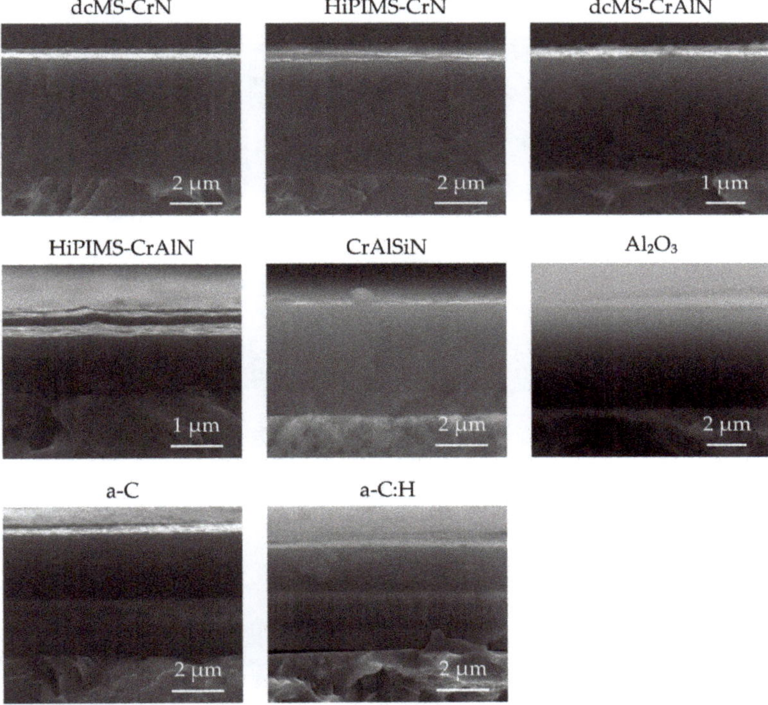

Figure 1. Scanning electron microscope (SEM) micrographs of the morphology of the physical vapor deposition (PVD) films.

The deposition parameters of the PVD films are summarized in Table 1. All the targets had a size of 500 mm × 88 mm. It is worthwhile to mention that the AlCr20 and AlCr24 targets consisted of aluminum (purity 99.5%) and 20 or 24 chromium plugs (purity 99.9%). An overview of the chemical composition and the mechanical properties of the eight PVD films used in this study is given in Table 2. In addition, the uncoated AISI H11 steel was used as reference for comparison with the PVD films.

Table 1. Deposition parameters of the PVD films.

Film System	dcMS- CrN	HiPIMS- CrN	dcMS- CrAlN	HiPIMS- CrAlN	CrAlSiN	Al$_2$O$_3$	a-C (Top Layer)	a-C:H (Top Layer)
Sputter mode	dcMS	HiPIMS	dcMS	HiPIMS	dcMS	mfMS	mfMS	mfMS
Target × cathode power (kW) or voltage (V)	2 Cr × 4.0 kW	2 Cr × 4.0 kW	2 AlCr20 × 5.0 kW	2 AlCr20 × 7.0 kW	2 AlCr24 × 5.0 kW 1 Cr × 1.0 kW 1 Si × 2.0 kW	2 Al × 410 V	2 C × 3 kW	2 C × 3 kW
Pulse frequency (Hz)	–	1000	–	600	–	–	–	–
Pulse duration (µs)	–	50	–	50	–	–	–	–
Mid frequency (kHz)	–	–	–	–	–	50	20	20
Pressure (mPa)	400	400	500	500	500	control	300	300
Ar flow (sccm)	300	300	120	80	120	300	control	control
Kr flow (sccm)	50	50	180	40	80	–	–	–
N$_2$ flow (sccm)	control	control	control	control	control	–	–	–
O$_2$ flow (sccm)	–	–	–	–	–	60	–	–
C$_2$H$_2$ flow (sccm)	–	–	–	–	–	–	–	35
Bias voltage (−V)	90	100	120	150	120	100	130	130
Deposition time (s)	10.000	20.000	21.000	20.000	21.200	7.200	18.000	12.500

dcMS—direct current magnetron sputtering; HiPIMS—high-power impulse magnetron sputtering; a-C—hydrogen-free amorphous carbon; a-C:H—hydrogenated amorphous carbon; control—controlled.

Table 2. Overview of the chemical composition, film thickness, hardness, and elastic modulus of the used PVD Films.

Film System	Chemical Composition [at.%]					Hardness [GPa]	Elastic Modulus [GPa]
	Cr	Al	Si	N	O		
dcMS-CrN	52.4 ± 0.8	–	–	47.6 ± 0.8	–	22.9 ± 1.3	301.8 ± 17.7
HiPIMS-CrN	64.0 ± 1.1	–	–	36.0 ± 1.1	–	24.6 ± 1.7	332.4 ± 15.8
dcMS-CrAlN	12.3 ± 0.5	36.0 ± 1.0	–	51.8 ± 1.5	–	26.7 ± 2.2	306.9 ± 14.4
HiPIMS-CrAlN	33.2 ± 1.0	14.5 ± 0.2	–	52.4 ± 1.2	–	33.3 ± 4.1	354.6 ± 36.7
dcMS-CrAlSiN	15.5 ± 0.6	24.3 ± 0.7	8.1 ± 0.2	52.2 ± 1.4	–	27.6 ± 1.7	291.5 ± 11.5
Al$_2$O$_3$	–	45.0 ± 0.2	–	–	55.0 ± 0.2	14.9 ± 0.9	193.4 ± 7.1
a-C	Hydrogen-free amorphous carbon					20.5 ± 1.5	190.7 ± 8.4
a-C:H	Hydrogenated amorphous carbon					16.4 ± 1.0	148.5 ± 6.1

The morphology and topography of the PVD films were investigated in SEM analysis using a FE-JSEM 7001 (Jeol, Akishima, Japan). In addition, the roughness profile of the PVD films and the uncoated AISI H11 was analyzed with the confocal 3D microscope μsurf (NanoFocus, Oberhausen, Germany). The arithmetical mean roughness, R_a, and the mean roughness depth, R_z, were determined according to DIN EN ISO 4287 and 4288 [20,21]. The contact angle measuring system G40 (Krüss, Hamburg, Germany) was employed, in order to characterize the surface free energy. Distilled water, ethylene glycol, dimethylformamide, 1-octanol as well as 1-decanol were used as test liquids. The disperse and polar components of their surface tension are given in Table 3. The static contact angles of five measuring points were determined for each surface system when wetted with the respective test liquids. The static contact angles were measured using an optical measuring system, which determined the drop profile and consequently calculated the angles. A total of 10 measurements per testing liquid and PVD film were carried out. The arithmetic mean value of the measured contact angle θ was used to calculate the disperse component, σ_s^D, and the polar component, σ_s^P, of the surface free energy, σ_s, according to the method by Owens, Wendt, Rabel, and Kaelble (OWRK) as described in DIN 55660-2 [22]. In this case, the values of the surface energy σ_l, with its disperse component σ_l^D and polar component σ_l^P, of the test liquids and the measured contact angle θ were used to calculate x and y according to Equations (1) and (2). These values were plotted in a x-y diagram in order to determine the regression line of the x and y values. Within this context, the polar component, σ_s^P, of the surface free energy corresponds to the square of the slope of the regression line, whereas the disperse component, σ_s^D, is the square of the ordinate intercept.

$$x = \sqrt{\frac{\sigma_l^P}{\sigma_l^D}} \tag{1}$$

$$y = \frac{(1 + \cos\theta) \times \sigma_l}{2\sqrt{\sigma_l^D}} \tag{2}$$

With the calculated values for the polar and dispersive component of the surface free energy of the different surface systems, the work of adhesion W_a between the surface systems and PA, PP, and PS was determined, based on the geometric mean equation for the work of adhesion:

$$W_a = 2\left(\sqrt{\sigma_s^D \times \sigma_b^D} + \sqrt{\sigma_s^P \times \sigma_b^P} \right). \tag{3}$$

The work of adhesion describes the theoretical work that is necessary to separate the polymer counterpart from the surface. A detailed description of the mathematical derivation of Equation (1) is given in [23]. The values of the disperse component, σ_b^D, and the polar component, σ_b^P, of PA, PP, and PS, as reported by Erhard, were representatively used for the calculations (see Table 4) [24]. The friction behavior of the PVD films and the uncoated AISI H11 was analyzed by employing a high-temperature tribometer with a ball-on-disc setup by Anton Paar (former: CSM Instruments, Buchs, Switzerland). Balls of the industrial relevant plastics Zytel® PA (DuPont, Midland, MI, USA), Pro-fax 6523 PP (LyondellBasell, Rotterdam, The Netherlands), and Crystal PS 1300 (Ineos, Lausanne, Switzerland) were used. The counterpart balls with a diameter of 6 mm were pressed with a load of $F_N = 10$ N and a constant sliding velocity of $v = 10$ cm/s against the surface systems for a total distance of 25 m. In order to recreate similar surrounding conditions as in the demolding process of the injection molding cycle, the temperature was set to the recommended tool temperature for the processing of the plastic. In the case of PP, PS, and PA, the temperature was set to 85, 60, and 40 °C, respectively. The friction force F_R was measured dynamically from start to end and the coefficients of friction were determined of each tribometer measurement. A total of three tribometer measurements were carried out for each testing condition in order to calculate the average value. After the tests, the volume of the balls was determined with the aid of light microscopic examinatisons in order to calculate the wear coefficient. A schematic representation of the experimental procedure is shown in Figure 2.

Table 3. Disperse and polar component of the surface energy of the test liquids.

Test Liquid	Surface Energy, σ_l [mN/m]	Disperse Component, σ_l^D [mN/m]	Polar Component, σ_l^P [mN/m]
Distilled water	72.8	21.8	51
Ethylene glycol	48	29	19
Dimethylformamide	37.3	32.4	4.9
1-Octanol	21.6	21.6	0
1-Decanol	28.5	22.2	6.3

Table 4. Disperse and polar component of the surface energy of the plastic counterparts according to Erhard [20].

Plastic Counterpart	Surface Energy, σ_b [mN/m]	Disperse Component, σ_b^D [mN/m]	Polar Component, σ_b^P [mN/m]
PA	47.5	36.8	10.7
PP	31.2	30.5	0.7
PS	42.0	41.4	0.6

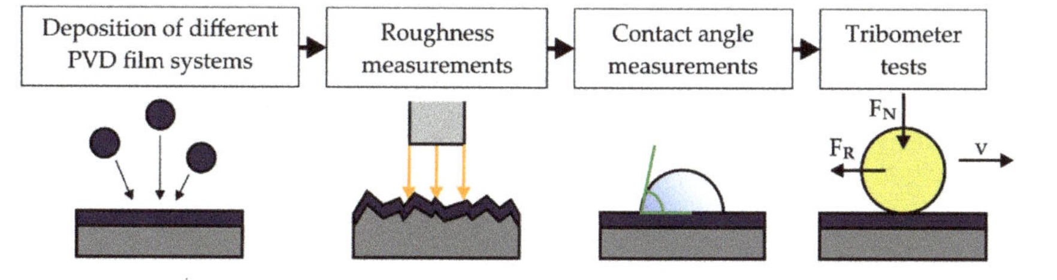

Figure 2. Schematic representation of the experimental procedure.

3. Results

3.1. Topography and Surface Roughness

The SEM micrographs of the topography show differences among the surface structure of the PVD films (see Figure 3). The CrN and CrAlN films sputtered in dcMS were marked by a cauliflower-like surface structure, with a smaller dimension in the case of CrAlN. This results from the smaller grain size of CrAlN in comparison to CrN [25,26]. In contrast, the HiPIMS variants of CrN and CrAlN exhibited a dense surface without any macroscopic growth defects, which is ascribed to the densification mechanism during the bombardment of highly ionized sputtered species [27]. A similar surface texture was observed for CrAlSiN, even though it was sputtered in dcMS. In this case, the denser surface results from the Si-doping, which induces the formation of a nanocomposite structure, consisting of nanoparticles embedded in an amorphous matrix [28]. In addition, it was observed that the HiPIMS-CrN, HiPIMS-CrAlN, and CrAlSiN films exhibited a saw-tooth-like surface structure, which is ascribed to the highly bombarded film growth. The Al_2O_3 film had an amorphous-like structure and also had a dense surface topography. In contrast, the topography of the carbon layers was distinguished by a cluster-like texture and was in good agreement with the observation of previous studies [29,30].

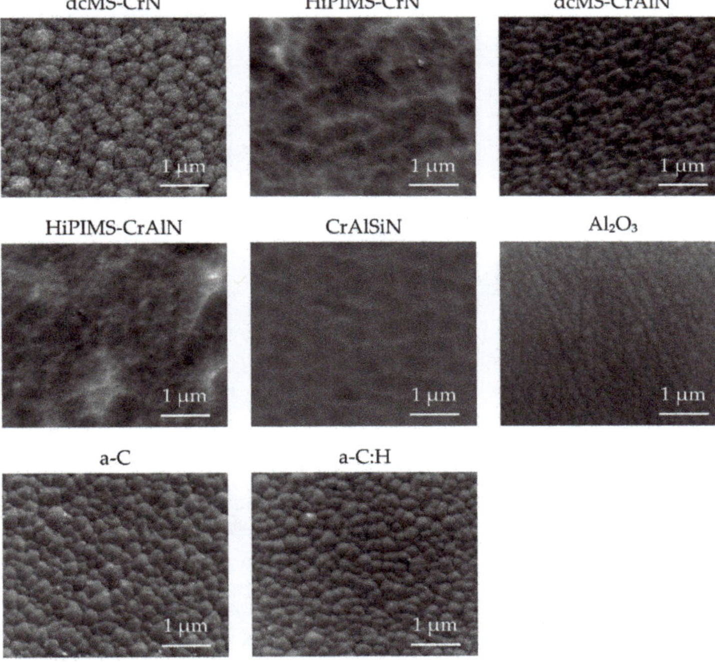

Figure 3. SEM micrographs of the topography of the PVD films.

The roughness profile of all surface systems was analyzed using the confocal 3D microscope. The measured arithmetical mean roughness values, R_a, as well as the mean roughness depth, R_z, are depicted in Figure 4. With exception of dcMS-CrN, it was noticeable that the measured values for R_a and R_z of the surface systems were comparable to each other and show similar tendencies. The uncoated AISI H11 steel had the lowest roughness by having values of $R_a = 4.4 \pm 0.3$ nm and $R_z = 34.8 \pm 4.4$ nm. For all PVD film systems, it was observed that the PVD deposition led to a roughness increase of the surfaces. The roughness increase is typical for depositions by PVD since the roughness asperities of the substrate disturb the trajectory of impinging material. As a consequence, the film growth is affected by the so-called shadow effect, which favors the formation of film growth defects and, hence, a higher roughness. Among the PVD films, the amorphous carbon films exhibited the lowest roughness values, which is typical for amorphous carbon films when compared to crystalline film systems. In contrast, the HiPIMS-CrN, HiPIMS-CrAlN, and CrAlSiN films showed the tendency of having the highest R_a and R_z values among the different PVD films. This behavior is attributed to the topography of these films, since they were marked by a saw-tooth-like structure due to the intensified bombardment during the film growth.

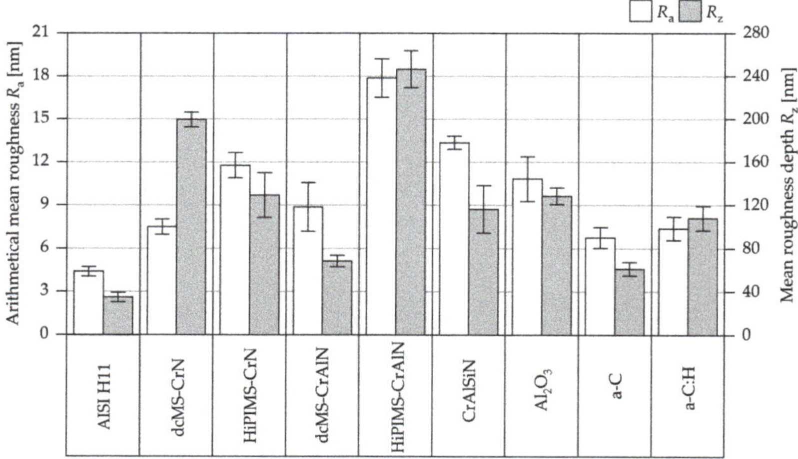

Figure 4. Arithmetical mean roughness, R_a, and mean roughness depth, R_z, of the AISI H11 steel and the PVD films.

3.2. Surface Free Energy and Work of Adhesion

In order to analyze the surface free energy of the different surface systems, the static contact angles were measured with five different testing liquids (see Table 5). It is noticeable that the standard deviation does not exceed 3° and thus fulfills the recommendation according DIN 55660-2 [22]. The disperse and polar components of the surface energies were calculated based on the results in accordance with the method by OWRK. The results are visualized in Figure 5. It is noticeable that all PVD films, with exception of the a-C:H film, are marked by lower free surface energies than the uncoated steel material with a value of 32.91 mN/m. Within this context, the disperse component of 13.83 mN/m was the lowest of all surface systems, whereas the polar amount of 14.58 mN/m was the highest value. The second highest polar surface energy was achieved by the a-C:H film with a value of 12.44 mN/m. The high polarity of the hydrogenated carbon layer was in good agreement with results reported by Bobzin et al., who determined a polar amount of 15 mN/m and a disperse component of 27.9 mN/m for a plasma assisted chemical vapor deposition (PACVD) deposited a-C:H film [31]. Moreover, the Al_2O_3 film also had a high value for the polar component with 7.53 mN/m, but the total surface free energy was 29.68 mN/m and, thus, was at the same level as the a-C layer with a value of 29.71 mN/m. The Cr-based nitrides were distinguished by surface energies in the range of 24.45 and 26.60 mN/m. Consequently, they had the lowest values among all surface systems. The polar fraction was, in particular, significantly lower in comparison to the other PVD films and the uncoated AISI H11. The values of the polar surface energy ranged between 1.18 and 2.48 mN/m. The dcMS-CrN and HiPIMS-CrN had surface energies of 25.63 and 24.45 mN/m, respectively. For an unspecified CrN film, Lugscheider et al. reported a surface free energy of approximately 38 mN/m, which is higher than the determined values for the present films [32]. However, the polar fraction of 2.48 and 1.18 mN/m of dcMS-CrN and HiPIMS-CrN are similar to the values determined by Lugscheider et al.

The surface energies of dcMS-CrAlN and HiPIMS-CrAlN were 26.60 and 25.30 mN/m, respectively, and are, hence, similar to values reported by Bobzin et al. [33]. In their investigation, the amount of Cr and Al was varied and values in the range between 25.64 and 29.07 mN/m were determined, but the polar content was significantly higher ranging from 5.44 to 6.79 mN/m. Moreover, they did not observe any dependence of the Al concentration on the surface free energy, which correlates with the surface energies of the Al-free CrN films and the CrAlN systems as well. However, the low surface energies of the Cr-based nitrides in contrast to the other PVD films can be ascribed to the chemical nature and is probably attributed to the nitrogen content. Theiss reported similar polar surface energies for CrAlN films with different chemical compositions, which were deposited by means of dcMS, mfMS, and HiPIMS as in this investigation [34]. However, the disperse components were considerably higher, so that values of up to 40 mN/m were obtained for the total surface free energy. Furthermore, Theiss investigated the surface free energy of AISI 420 steel and observed a reduced polar component. According to X-ray photoelectron spectroscopy (XPS) measurements, this behavior was ascribed to the formation of oxides and carbon bondings in the surface reaction layer as well as of adsorbates on the surface [34].

Table 5. Arithmetical mean of the measured static contact angles.

Surface System	Static Contact Angle [°]				
	Water	Ethylene Glycol	Dimethylformamide	1-Octanol	1-Decanol
Uncoated AISI H11	69.4 ± 1.7	59.6 ± 2.07	32.6 ± 2.1	2.0 ± 2.1	14.4 ± 1.8
dcMS-CrN	93.2 ± 0.8	77.2 ± 0.8	35.2 ± 1.6	11.8 ± 0.8	21.8 ± 1.8
HiPIMS-CrN	100.2 ± 0.8	75.6 ± 1.1	50.4 ± 1.7	13.2 ± 1.5	26.4 ± 2.5
dcMS-CrAlN	94.6 ± 1.1	75.0 ± 2.6	34.2 ± 1.9	0	9.6 ± 2.1
HiPIMS-CrAlN	94.6 ± 1.3	75.2 ± 0.8	42.0 ± 1.6	4.8 ± 1.9	24.4 ± 1.5
CrAlSiN	94.4 ± 1.8	72.8 ± 1.9	41.4 ± 1.7	7.2 ± 1.8	25.4 ± 2.3
Al_2O_3	81.2 ± 1.6	59.6 ± 1.1	28.4 ± 1.1	0	12.8 ± 0.8
a-C	88.2 ± 1.8	58.8 ± 2.3	15.4 ± 1.7	0	0.8 ± 1.1
a-C:H	73.2 ± 1.8	39.4 ± 0.9	10.4 ± 1.5	0	1.8 ± 1.8

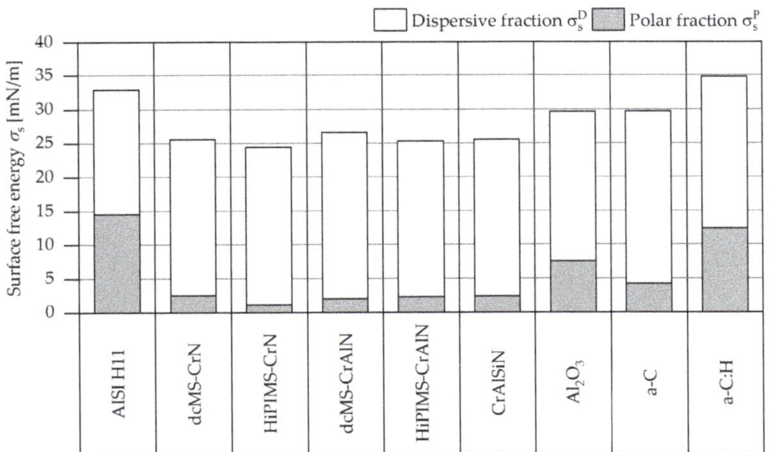

Figure 5. Surface free energy, σ_s, with corresponding disperse and polar components of AISI H11 steel and PVD films.

In order to evaluate the effect of the surface energies on the adhesion behavior of the surface systems with the plastic counterparts, the work of adhesion, W_a, was determined using Equation (1).

The work of adhesion of the AISI H11 steel and PVD films is illustrated in Figure 6. For the calculations, the values of disperse and polar surface energies of PA, PP, and PS, were selected according to Erhard (see Table 4) [24]. It is noticeable that both PP and PS have the lowest work of adhesion to the uncoated material. This behavior is ascribed to the fact that these materials are non-polar plastics with a very low fraction of polar surface energy. As the uncoated AISI H11 steel has a lower disperse component than the PVD films, only the disperse surface energy has an impact on the adhesion behavior to non-polar plastics since only a very low proportion of polar bonds can be formed. In the case of polar plastics such as PA, the polar component of the surface energy is more relevant for the adhesion behavior, so that surface systems with a low polar surface energy are marked by a low work of adhesion. In particular, the Cr-based nitrides have the lowest values of polar surface energy and, thus, exhibit the lowest work of adhesion to PA by having values ranging from 65.6 to 69.4 mN/m. These results are in good agreement with the theoretical results of Lugscheider, who ascribes the adhesion behavior of solid surfaces to interactions between polar and disperse surface energies [32]. As a consequence, a non-polar surface is only influenced by the disperse fraction. Moreover, the calculated results were consistent with observations made by Theiss. In his work, the work of adhesion of PE and PP was not reduced by CrAlN films when compared to the uncoated AISI 420 steel [34]. However, a significant reduction of the adhesion was reported for the polar polymethyl methacrylate (PMMA). With regard to the friction behavior, it is expected that the work of adhesion of the surface systems to the plastic counterpart will have an impact on the adhesion component of the friction.

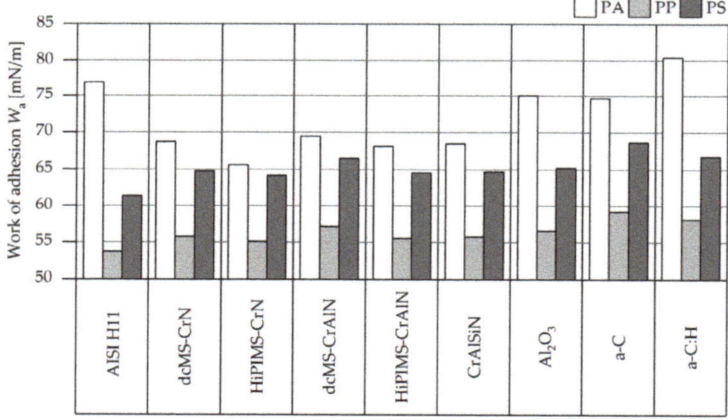

Figure 6. Work of adhesion, W_a, of the AISI H11 steel and PVD films with PA, polypropylene (PP), and polystyrene (PS).

3.3. Tribological Properties

The friction behavior of the uncoated steel and PVD films was analyzed by means of tribometer tests in a ball-on-disc setup. The determined coefficients of friction are illustrated in Figure 7. With exception of the uncoated steel, it is noticeable that PS leads to the highest friction for all PVD films when compared to PA and PP. This behavior is in agreement with the values determined by Hellerich et al., who reported a similar friction behavior of the three plastic counterparts against steel, even though the steel surface had a roughness, R_z, of 2 µm, which is, hence, higher than the roughness profile of the PVD films [35]. When sliding against PA and PS, the uncoated steel material had higher coefficients of friction of 0.78 ± 0.05 and 0.72 ± 0.03 than the PVD films. In the case of PP, the steel material was also distinguished by a higher friction when compared to most PVD films. The Al$_2$O$_3$ film exhibited relatively high coefficients of friction 0.55 ± 0.03, 0.53 ± 0.04, and 0.58 ± 0.03 for PA, PP, and PS, respectively. This high friction for uncoated steel is ascribed to the low roughness values, which lead to an increase of the adhesive component of friction and, thus, to higher coefficients of friction [36]. Furthermore, a higher coefficient of friction against PA when compared to dcMS-CrAlN was observed for the HiPIMS-CrAlN film. However, the work of adhesion was slightly lower for HiPIMS-CrAlN than for dcMS-CrAlN, so that the topography of the surfaces further affects the friction behavior, since the HiPIMS variant has a higher roughness. The friction against plastic components mainly consists of a deformative and adhesive component. Therefore, it is also necessary to analyze the influence of the roughness profile of the surface systems on the coefficient of friction.

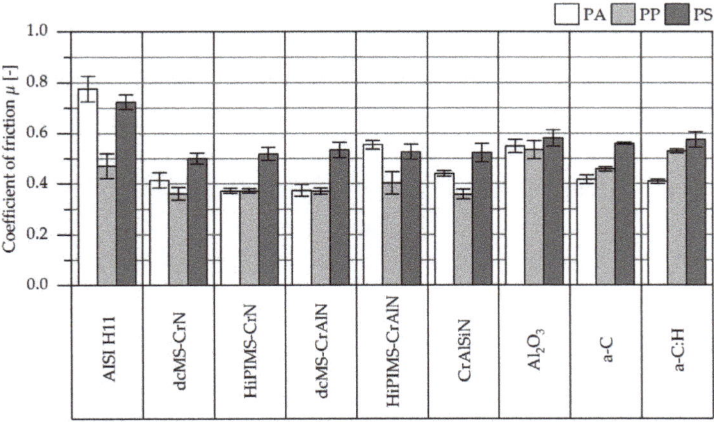

Figure 7. Coefficients of friction of the AISI H11 steel and PVD films.

The coefficient of friction was plotted against the mean roughness depth, R_z, as illustrated in Figure 8 for a better understanding of the correlation between friction behavior and roughness profile. The σ confidence band and σ prediction band of the regression line were also included. The tendency of a slightly higher friction was observed for the surface systems with a higher surface roughness. A similar trend was observed when plotting against the arithmetic mean roughness, R_a. This behavior was particularly noted for PS. However, it is worthwhile mentioning that some PVD systems do not fully follow this trend for PA and PP by being outside the range of the regression line. This was particularly noted for PA, which is characterized by a broad range of the σ confidence band and σ prediction band. It is concluded that the friction behavior of the surface systems is essentially influenced by the adhesive mechanisms. The adhesive friction component decreases with an increasing surface roughness, whereas the deformative component increases steadily [37]. However, the deformative friction component is not significant in this roughness region as the coefficient of friction does not significantly increase with higher roughness values of the surfaces. Therefore, the adhesive component is the dominant factor for the friction behavior of the PVD films.

As the adhesive component was identified as decisive for the friction behavior, the coefficient of friction was also plotted against the work of adhesion for PA, PP, and PS (see Figure 9). In case of PA and PP, it was observed that a higher work of adhesion leads to higher coefficients of friction, which is ascribed to the increase of the adhesive friction component. This behavior is in good agreement with the study by Bagcivan et al., who investigated the adhesion and tribological behavior of CrAlON films against Plexiglas [12]. They noted lower coefficients of friction for CrAlON films with lower adhesion against Plexiglas. However, this trend was not seen for PS as an opposite relation was observed. In this case, the friction decreases with an increasing work of adhesion. In order to understand this phenomenon, the wear behavior of the PA, PP, and PS balls was investigated by determining the wear coefficient. The determined wear coefficients of the balls are given in Figure 10. It is noticeable that the PS balls are exposed to a higher wear in the tribometer tests than the PA and PP balls. In the tribometer tests, an agglomeration of wear debris from the PS balls along the wear track was observed. It is assumed that the PS wear debris had a friction-reducing effect and resulted in a lower coefficient of friction despite the higher work of adhesion. This behavior was reported by Thorp, who observed the formation of a transfer film of polytetrafluoroethylene (PTFE) and ultra-high molecular weight polyethylene (UHMWPE) on the steel surfaces during dry sliding, which leads to reduced coefficients of friction [38]. A similar mechanism is expected, as adhered particles were observed on the surface, after the tribometer tests with PS balls, whereas no material transfer was observed for PA and PP. However, the uncoated surface is an exception, which does not cause a high wear of the PS balls. In this case, the wear coefficient of the PS balls was comparable to the wear coefficients of PA and PP. This behavior is ascribed to several reasons, which are related to the roughness profile and to the work of adhesion of the uncoated AISI H11. On the one hand, the uncoated steel material had a mean roughness depth, R_z, of 34.8 ± 4.4 nm and a low hardness of 7.03 ± 0.31 GPa, which in combination do not lead to abrasive wear mechanisms in contrast to the rougher and harder PVD films. On the other hand, the work of adhesion for both the AISI H11 and PS was 61.42 mN/m and thereby significantly lower in comparison to the PVD films. The adhesion strength was therefore too low to induce adhesively induced wear.

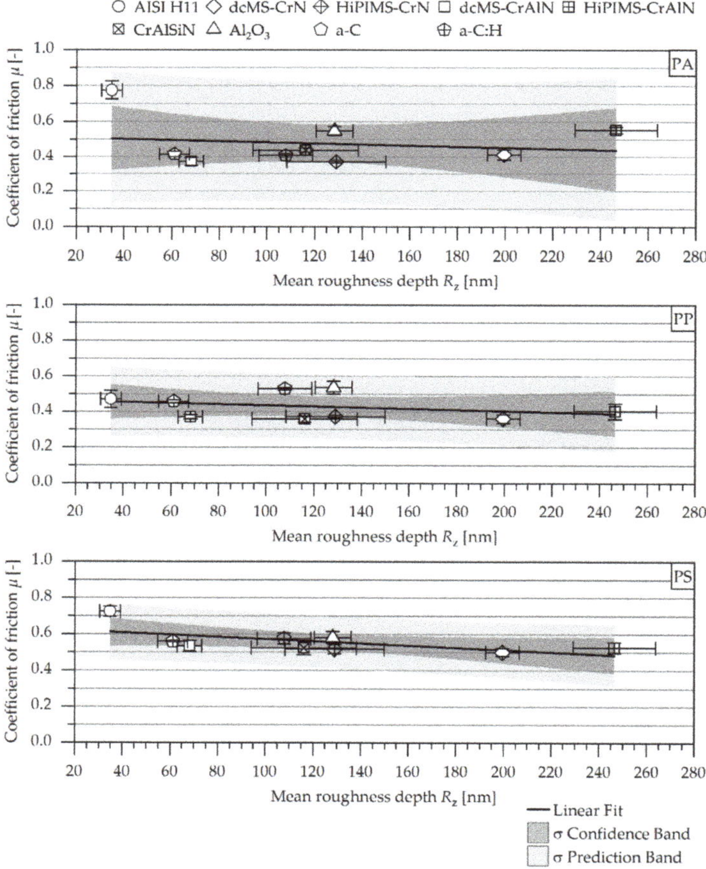

Figure 8. Correlation between mean roughness depth, R_z, and the coefficient of friction of the AISI H11 steel and the PVD films against PA, PP, and PS.

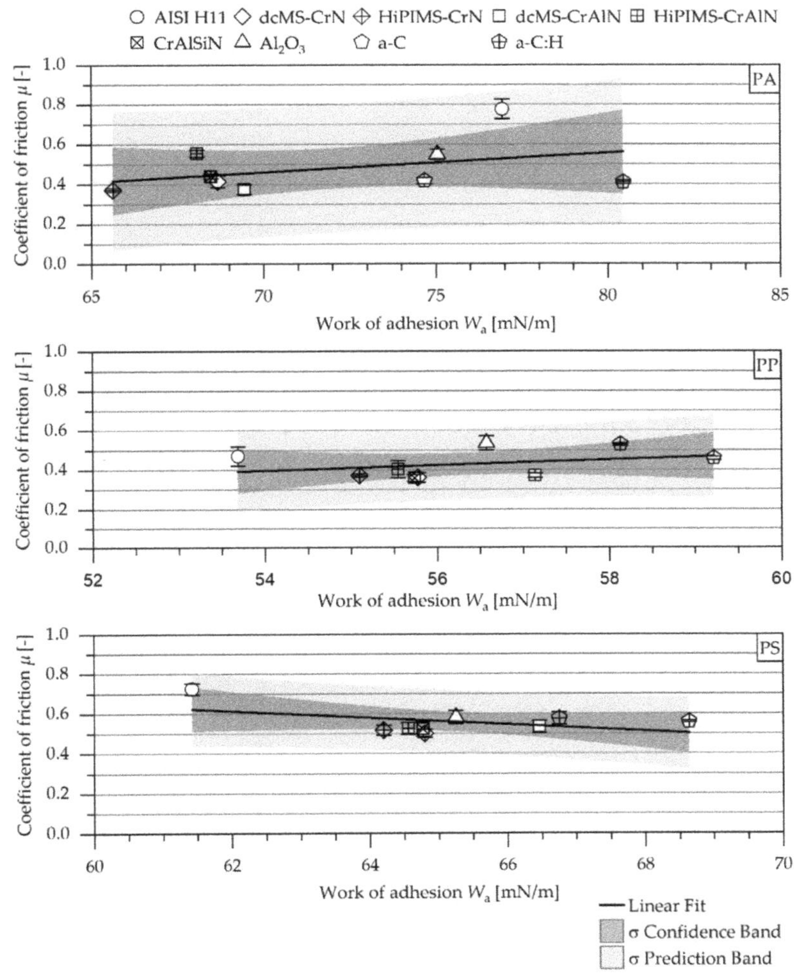

Figure 9. Correlation between work of adhesion W_a and the coefficient of friction of the AISI H11 steel and PVD films against PA, PP, and PS.

The curves of the coefficient of friction are exemplarily shown in Figure 11 for the uncoated AISI H11 and the dcMS-CrN film against PA, PP, and PS for a comprehensive understanding of the wear mechanisms of the plastic balls. When comparing the friction behavior of the uncoated steel and the dcMS-CrN film against PS, it is noticeable that the curve for dcMS-CrN is marked by a strong stick–slip effect, which is the explanation for the strong wear of the PS balls. Moreover, the stick–slip effect increases with larger sliding distance due to an increased wear volume of the PS balls. In contrast, the curves for the tribometer tests with PA and PS remain almost constant with an increasing sliding distance. This can be ascribed to the low wear of the PA and PS balls. Regarding the wear behavior, it is worth mentioning that the surface systems did not show any sign of abrasive wear, a fact that is attributed to the high hardness difference between the friction partners. Only an adhesive material transfer of the worn plastic balls was observed on the surfaces. The dimensions of the transfer depend on the wear coefficients of the balls.

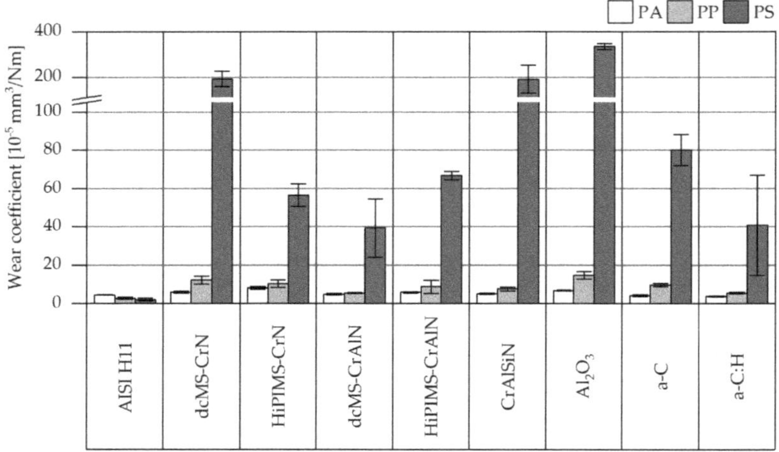

Figure 10. Wear coefficients of the PA, PP, and PS balls after sliding against the AISI H11 steel and PVD films.

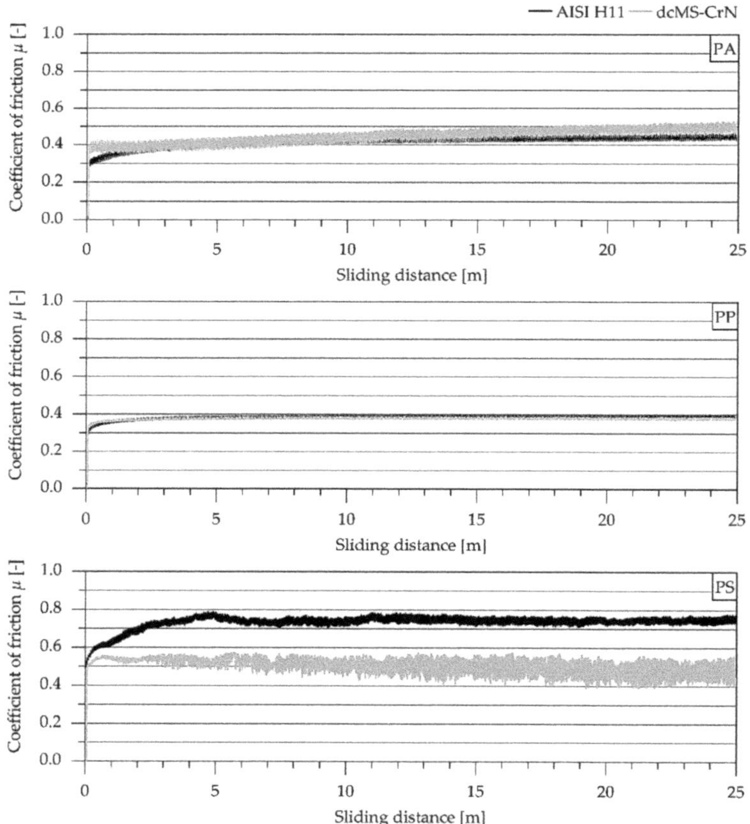

Figure 11. Exemplary curves of the coefficients of friction of the AISI H11 steel and dcMS-CrN against PA, PP, and PS.

4. Conclusions

The cause–effects relationship between the surface roughness and surface free energy of PVD films on the friction behavior against PA, PP, and PS was investigated. Besides the uncoated AISI H11 steel, several film systems, consisting of dcMS- and HiPIMS-sputtered CrN and CrAlN, as well as CrAlSiN, Al_2O_3, a-C, and a-C:H, were systematically analyzed with regard to their surface free energy and tribological properties. Among the surface systems, differences concerning the surface free energy and, in particular, the polar component, were observed. The Cr-based nitrides exhibit the lowest surface energies with a very low polar content. Based on the results of the surface free energy, the work of adhesion was calculated for PA, PP, and PS. It was shown that the disperse component of the surface systems is decisive for the work of adhesion with the non-polar PP and PS, since no polar interactions occur on the surface. In contrast to the polar PA, the polar component is more significant than the dispersive component, so that the Cr-based nitrides with low polar surface energy are marked by the lowest values. In tribometer tests, an influence of the surface roughness and the work of adhesion on the friction behavior was identified. The trend of lower coefficients of friction with slightly higher roughness was observed, in particular for PS. In contrast, a higher work of adhesion results in higher coefficients of friction when sliding against PA and PP. However, a contrary relationship was observed in case of PS, which is ascribed to the formation of a transfer film due to the high wear of PS balls. Therefore, a dependence of both properties, the roughness and the work of adhesion, on the friction behavior of PVD films against the plastic counterparts was shown.

Author Contributions: Conceptualization, W.T., N.F.L.D., D.S. and N.G; Methodology, N.F.L.D., D.S. and N.G.; Investigation, N.F.L.D., D.S. and N.G.; Writing—Original Draft, N.F.L.D. and N.G.; Writing—Review and Editing, W.T. and D.S.; Visualization, N.F.L.D. and N.G.; Supervision, W.T.

Funding: This research received no external funding.

Acknowledgments: The authors acknowledge the financial support by the German Research Foundation and TU Dortmund University within the funding program "Open Access Publishing". The authors wish to express their gratitude to Dirk Biermann and Eugen Krebs from the Institute of Machining Technology (TU Dortmund University) for providing the confocal 3D microscope and Jörg C. Tiller and Christian Krumm of the Biomaterials and Polymer Science Department (TU Dortmund University) for providing the contact angle measuring system.

Conflicts of Interest: The authors declare no conflict of interest.

References

1. Zheng, R.; Tanner, R.I.; Fan, X.-J. *Injection Molding. Integration of Theory and Modeling Methods*; Springer Science & Business Media: Berlin, Germany, 2011; ISBN 978-3-642-21262-8.
2. Gramann, P.J.; Osswald, T.A. Introduction. In *Injection Molding Handbook*, 2nd ed.; Osswald, T.A., Turng, L.-S., Gramann, P.J., Eds.; Springer Science & Business Media: Berlin, Germany, 2008; pp. 1–18. ISBN 978-3-446-40781-7.
3. Rosato, D.V.; Rosato, D.V.; Rosato, M.G. *Injection Molding Handbook*, 3rd ed.; Springer US: Boston, MA, USA, 2000; ISBN 9781461370772.
4. Kerkstra, R.; Brammer, S. *Injection Molding Advanced Troubleshooting Guide*; Carl Hanser Verlag: Munich, Germany, 2018; ISBN 9781569906453.
5. Kazmer, D. *Injection Mold Design Engineering*, 2nd ed.; Carl Hanser Verlag: Munich, Germany, 2016; ISBN 978-1-56990-570-8.
6. Sakai, T.; Kikugawa, K. Injection Molding Machinery and Systems. In *Injection Molding: Technology and Fundamentals*; Kamal, M.R., Agassant, J.-F., Eds.; Carl Hanser Verlag: Munich, Germany, 2009; pp. 73–131. ISBN 978-3-446-41685-7.
7. Bienk, E.J.; Mikkelsen, N.J. Application of advanced surface treatment technologies in the modern plastics moulding industry. *Wear* **1997**, *207*, 6–9. [CrossRef]
8. Bobzin, K.; Nickel, R.; Bagcivan, N.; Manz, F.D. PVD—Coatings in Injection Molding Machines for Processing Optical Polymers. *Plasma Process. Polym.* **2007**, *4*, S144–S149. [CrossRef]

9. Silva, F.J.G.; Martinho, R.P.; Baptista, A.P.M. Characterization of laboratory and industrial CrN/CrCN/diamond-like carbon coatings. *Thin Solid Films* **2014**, *550*, 278–284. [CrossRef]

10. Silva, F.; Martinho, R.; Andrade, M.; Baptista, A.; Alexandre, R. Improving the wear resistance of moulds for the injection of glass fibre–reinforced plastics using PVD coatings: A comparative study. *Coatings* **2017**, *7*, 28. [CrossRef]

11. D'Avico, L.; Beltrami, R.; Lecis, N.; Trasatti, S. Corrosion behavior and surface properties of PVD coatings for mold technology applications. *Coatings* **2019**, *9*, 7. [CrossRef]

12. Bagcivan, N.; Bobzin, K.; Brögelmann, T.; Kalscheuer, C. Development of (Cr,Al) ON coatings using middle frequency magnetron sputtering and investigations on tribological behavior against polymers. *Surface Coat. Technol.* **2014**, *260*, 347–361. [CrossRef]

13. Bobzin, K.; Grundmeier, G.; Brögelmann, T.; los Arcos, T.; Wiesing, M.; Kruppe, N.C. Nitridische und oxinitridische HPPMS-Beschichtungen für den Einsatz in der Kunststoffverarbeitung (Teil 2). *Vakuum Forschung Praxis* **2017**, *29*, 24–28. [CrossRef]

14. *Paints and Varnishes-Pull-off Test for Adhesion (ISO 4624:2016), German Version EN ISO 4624:2016*; DIN German Institute for Standardization: Berlin, Germany, 2016.

15. Sasaki, T.; Koga, N.; Shirai, K.; Kobayashi, Y.; Toyoshima, A. An experimental study on ejection forces of injection molding. *Precis. Eng.* **2000**, *24*, 270–273. [CrossRef]

16. Burkard, E.; Walther, T.; Schinköthe, W. Influence of mold wall coatings while demoulding in the injection molding process. *Stuttg. Kunststoff Kolloquium* **1999**, *16*, 1–8.

17. Kobayashi, Y.; Shirai, K.; Sasaki, T.; Kobayashi, Y.; Shirai, K.; Sasaki, T. Relationship between core surface roughness and ejection force for injection molding. *J. Jpn. Soc. Prec. Eng.* **2001**, *67*, 510–514. [CrossRef]

18. Mitschang, P.; Schledjewski, R.; Schlarb, A.K. Molds for continuos fibre reinforced polymer composites. In *Mold-Making Handbook*, 3rd ed.; Mennig, G., Stoeckhert, K., Eds.; Carl Hanser Verlag: Munich, Germany, 2013; pp. 200–238. ISBN 978-1-56990-446-6.

19. Hoffmann, F. Beitrag zur Charakterisierung des tribologischen Verhaltens von Diamantähnlichen Kohlenstoffschichten für die Holzbearbeitung. Ph.D. Thesis, TU Dortmund University, Dortmund, Germany, 2012.

20. *Geometrical Product Specifications (GPS)–Surface Texture: Profile Method–Terms, Definitions and Surface Texture Parameters (ISO 4287:1997 + Cor 1:1998 + Cor 2:2005 + Amd 1:2009), German Version EN ISO 4287:1998 + AC:2008 + A1:2009*; DIN German Institute for Standardization: Berlin, Germany, 2010.

21. *Geometrical Product Specifications (GPS)-Surface Texture; Profile Method-Rules and Procedures for the Assessment of Surface Texture (ISO 4288:1996), German Version EN ISO 4288:1997*; DIN German Institute for Standardization: Berlin, Germany, 1998.

22. *Paints and Varnishes–Wettability–Part 2: Determination of the Free Surface Energy of Solid Surfaces by Measuring the Contact Angle, German Version DIN 55660-2*; DIN German Institute for Standardization: Berlin, Germany, 2011.

23. Zhang, J. Work of adhesion and work of cohesion. In *Encyclopedia of Tribology*; Wang, Q.J., Chung, Y.-W., Eds.; Springer US: Boston, MA, USA, 2013; pp. 4127–4132. ISBN 978-0-387-92896-8.

24. Erhard, G. *Designing with Plastics*; Carl Hanser Verlag: Munich, Germany, 2006; ISBN 978-3-446-22590-9.

25. Barshilia, H.C.; Selvakumar, N.; Deepthi, B.; Rajam, K.S. A comparative study of reactive direct current magnetron sputtered CrAlN and CrN coatings. *Surf. Coat Technol.* **2006**, *201*, 2193–2201. [CrossRef]

26. Lin, J.; Mishra, B.; Moore, J.J.; Sproul, W.D. Microstructure, mechanical and tribological properties of Cr1−xAlxN films deposited by pulsed-closed field unbalanced magnetron sputtering (P-CFUBMS). *Surf. Coat Technol.* **2006**, *201*, 4329–4334. [CrossRef]

27. Ehiasarian, A.; Münz, W.D.; Hultman, L.; Helmersson, U.; Petrov, I. High power pulsed magnetron sputtered CrNx films. *Surf. Coat Technol.* **2003**, *163–164*, 267–272. [CrossRef]

28. Rafaja, D.; Dopita, M.; Růžička, M.; Klemm, V.; Heger, D.; Schreiber, G.; Šímac, M. Microstructure development in Cr–Al–Si–N nanocomposites deposited by cathodic arc evaporation. *Surf. Coat Technol.* **2006**, *201*, 2835–2843. [CrossRef]

29. Tillmann, W.; Lopes Dias, N.F.; Stangier, D. Influence of plasma nitriding pretreatments on the tribo-mechanical properties of DLC coatings sputtered on AISI H11. *Surface Coat. Technol.* **2019**, *357*, 1027–1036. [CrossRef]

30. Tillmann, W.; Lopes Dias, N.F.; Stangier, D.; Maus-Friedrichs, W.; Gustus, R.; Thomann, C.A.; Moldenhauer, H.; Debus, J. Improved adhesion of a-C and a-C: H films with a CrC interlayer on 16MnCr5 by HiPIMS-pretreatment. *Surface Coat. Technol.* **2019**, *375*, 877–887. [CrossRef]

31. Bobzin, K.; Bagcivan, N.; Theiss, S.; Yilmaz, K. Plasma coatings CrAlN and a-C: H for high efficient power train in automobile. *Surface Coat. Technol.* **2010**, *205*, 1502–1507. [CrossRef]
32. Lugscheider, E.; Bobzin, K. The influence on surface free energy of PVD-coatings. *Surface Coat. Technol.* **2001**, *142*, 755–760. [CrossRef]
33. Bobzin, K. Benetzungs- und Korrosionsverhalten von PVD-beschichteten Werkstoffen für den Einsatz in umweltverträglichen Tribosystemen. Ph.D. Thesis, RWTH Aachen, Aachen, Germany, 2000.
34. Theiss, S. Analyse gepulster Hochleistungsplasmen zur Entwicklung neuartiger PVD-Beschichtungen für die Kunststoffverarbeitung. Ph.D. Thesis, RWTH Aachen, Aachen, Germany, 2013.
35. Hellerich, W.; Harsch, G.; Haenle, S. *Werkstoff-Führer Kunststoffe. Eigenschaften, Prüfungen, Kennwerte*; Carl Hanser Verlag: Munich, Germany, 2010; ISBN 978-3-446-42436-4.
36. Goryacheva, I.G. *Contact Mechanics in Tribology*; Springer Science & Business Media: Berlin, Germany, 2011; ISBN 978-90-481-5102-8.
37. Brinksmeier, E.; Riemer, O.; Twardy, S. Tribological behavior of micro structured surfaces for micro forming tools. *Int. J. Mach. Tools Manuf.* **2010**, *50*, 425–430. [CrossRef]
38. Thorp, J.M. Tribological properties of selected polymeric matrix composites against steel surfaces. In *Friction and Wear of Polymer Composites*; Friedrich, K., Ed.; Elsevier Science: Amsterdam, The Netherlands, 1986; pp. 89–135. ISBN 9780444597113.

 *coatings*

Article

Two Different Scenarios for the Equilibration of Polycation—Anionic Solutions at Water–Vapor Interfaces

Eduardo Guzmán [1,2,*], Laura Fernández-Peña [1], Andrew Akanno [1,2], Sara Llamas [1], Francisco Ortega [1,2] and Ramón G. Rubio [1,2]

[1] Departamento de Química Física, Facultad de Ciencias, Universidad Complutense de Madrid, Ciudad Universitaria s/n, 28040 Madrid, Spain
[2] Instituto Pluridisciplinar, Universidad Complutense de Madrid, Paseo Juan XXIII, 1, 28040 Madrid, Spain
* Correspondence: eduardogs@quim.ucm.es; Tel.: +34-91-394-4107

Received: 24 June 2019; Accepted: 11 July 2019; Published: 13 July 2019

Abstract: The assembly in solution of the cationic polymer poly(diallyldimethylammonium chloride) (PDADMAC) and two different anionic surfactants, sodium lauryl ether sulfate (SLES) and sodium N-lauroyl-N-methyltaurate (SLMT), has been studied. Additionally, the adsorption of the formed complexes at the water–vapor interface have been measured to try to shed light on the complex physico-chemical behavior of these systems under conditions close to that used in commercial products. The results show that, independently of the type of surfactant, polyelectrolyte-surfactant interactions lead to the formation of kinetically trapped aggregates in solution. Such aggregates drive the solution to phase separation, even though the complexes should remain undercharged along the whole range of explored compositions. Despite the similarities in the bulk behavior, the equilibration of the interfacial layers formed upon adsorption of kinetically trapped aggregates at the water–vapor interface follows different mechanisms. This was pointed out by surface tension and interfacial dilational rheology measurements, which showed different equilibration mechanisms of the interfacial layer depending on the nature of the surfactant: (i) formation layers with intact aggregates in the PDADMAC-SLMT system, and (ii) dissociation and spreading of kinetically trapped aggregates after their incorporation at the fluid interface for the PDADMAC-SLES one. This evidences the critical impact of the chemical nature of the surfactant in the interfacial properties of these systems. It is expected that this work may contribute to the understanding of the complex interactions involved in this type of system to exploit its behavior for technological purposes.

Keywords: polyelectrolyte; surfactants; kinetically trapped aggregates; interfaces; surface tension; interfacial dilational rheology; adsorption

1. Introduction

The study of polyelectrolyte oppositely charged surfactant solutions, either in bulk or close to interfaces (fluid and solid ones), has grown very fast in the last two decades [1], mainly as result of its interest for a broad range of technological and industrial fields, e.g., drug delivery systems, food science, tertiary oil recovery, or cosmetic formulations [1–9]. Most of such applications take advantage of the chemical nature of the compounds involved, structural features of the formed complexes, and the rich phase diagrams appearing in this type of system [10–12].

Despite the extensive research, the description of the physico-chemical behavior of these colloidal systems remains controversial, in part because the self-assembly processes of polyelectrolytes and surfactants bearing opposite charges leads to the formation of non-equilibrium complexes [10,13–16]. They are expected to impact significantly on the properties of the solutions and in their adsorption at the interfaces [7]. This makes it necessary to pay attention to aspects such as the polymer-surfactant mixing protocol, the elapsed time from the preparation of solutions until their study, or the addition of inert electrolytes when comparisons between different studies are performed [17–19]. The role of the above-mentioned aspects in the physico-chemical properties and the phase diagrams of polyelectrolyte-surfactant solutions have been the focus of many studies, which have evidenced the complex behavior of polyelectrolyte-surfactant solutions [6,8,17–19]. It is worth mentioning that the non-equilibrium nature of the complexation process of polymer-surfactant solutions has an extraordinary impact on the interfacial properties of such solutions, as was recently stated by Campbell and Varga [20].

The role of the presence of non-equilibrium aggregates on the adsorption of polymer-surfactant solutions at fluid interfaces was already evidenced by the seminal works of the groups of Campbell and of Meszaros, focused on the analysis of the surface tension of polyelectrolyte-surfactant solutions [7,21–23]. However, it was necessary to use neutron reflectometry, which provides information on the composition and structure of the interfaces to deepen the most fundamental aspects of the physico-chemical behavior of these systems [24–26]. The studies of Penfold's group were a preliminary step toward the understanding of the correlations existing between the aggregation occurring in polyelectrolyte-surfactant solutions and the behavior of these complexes' fluid interfaces [27–31]. However, such works used an extended Gibbs formalism to describe the adsorption at fluid interfaces, i.e., provide a thermodynamic description. This approach was able to account for the non-regular dependences of the surface tension on the bulk concentration (surface tension peaks), even though it neglects the impact of non-equilibrium aspects [32,33]. More recently, Campbell et al. [17,18,34–39], using surface tension measurements and neutron reflectometry combined with ellipsometry, Brewster angle microscopy, and different bulk characterization techniques, tried to link the interfacial properties of the solutions to the bulk phase behavior, paying special attention to the role of the non-equilibrium effects. Their physical picture takes into account the role of the depletion of the interface as a result of the aggregation in the bulk [40], and the enrichment of the interface in virtue of direct interactions of the formed aggregates [19].

Most studies that analyze the behavior of the adsorption of polyelectrolyte-surfactant solutions at fluid interfaces only consider the interfaces as static systems. However, a comprehensive description of their behavior requires taking into consideration the response of such systems against mechanical deformations, i.e., the rheological response of the interfaces [7,41–45]. The understanding of such aspects is essential because most technological applications of interfacial systems, e.g., foam stabilization [42], rely on the response of the interfaces against mechanical perturbations [43]. The seminal studies on the rheological characterization of polyelectrolyte-surfactant layers at the water–vapor interface done by Regismond et al. [26,46] pointed out the strong synergetic effect on the interfacial properties as result of the influence of the bulk complexation process in the interfacial properties. More recent studies by Bhattacharyya et al. [47] and Monteux et al. [48] correlated the interfacial rheological response of polyelectrolyte-surfactant solutions with their ability to stabilize foams. They found that the formation of gel-like layers at the interface hindered destabilization processes such as bubble coalescence and foam drainage. Deepening the understanding of the rheological response of polyelectrolyte-surfactant solutions, Noskov et al. [26,42,43,45,49] showed that the mechanical behavior of the interface is controlled by the heterogeneity of layers, which is reminiscent of the structure of the complexes formed in solution.

It is worth mentioning that most studies in the recent literature deal with solutions containing relative low polymer concentrations, which hold limited interest from an industrial point of view. It is expected that polymer concentration can present an important contribution in both the complexation process and the interfacial properties of polyelectrolyte-surfactant solutions [19,41,43]. Previous studies have shown that, whereas in diluted polyelectrolyte-surfactant solutions, equilibrium between free surfactant molecules and complexes is always present in solution, the role of the free surfactant is rather limited when polymer concentration is increased. For the latter, the binding degree of surfactant molecules to the polymer chain reach values above 90%, which makes it possible to assume that they are mostly complexes that are presented in solution, even for compositions in the vicinity of the onset of the phase separation region [50]. The differences in the complexation phenomena occurring in concentrated and diluted mixtures may significantly affect the interfacial assembly of polymer-surfactant solutions, with concentrated mixtures leading to the formation of interfacial layers, with composition mirroring the composition of the bulk solutions. The latter is far from the scenario found for diluted solutions [50,51].

This work presents a comparative study of the equilibrium and dynamic properties of interfacial layers formed upon adsorption at the water–vapor interface of solution formed by poly(diallyldimethylammonium chloride) (PDADMAC) and two different anionic surfactants: sodium lauryl-ether sulfate (SLES) and sodium N-lauroyl-N-methyltaurate (SLMT). PDADMAC was chosen as the polymer because of its common utilization as a conditioner in cosmetic formulations for hair care and cleansing. Furthermore, SLES and SLMT have been recently included in formulations of shampoos to replace sodium dodecylsulfate (SDS) due to their softness and mildness, which limits skin and mucosa irritation [1].

The main aim of this work is to unravel the different interfacial behavior appearing in polycation-oppositely charged surfactant mixtures. The adsorption at the water–vapor interface is studied by surface tension measurements obtained with different tensiometers. It is worth mentioning that although polyelectrolyte-surfactant may be out of equilibrium, for simplicity we will refer to the effective property measured in this work as surface tension. In addition to the steady state measurements of the surface tension, we will follow the adsorption kinetics of the complexes at the water–vapor interface by the time evolution of the surface tension (dynamic surface tension) and the mechanical performance of the interfaces against dilation using oscillatory barrier experiments in a Langmuir trough [52]. The obtained results will be combined with the information obtained from the study of the self-assembly phenomena taking place in solution. This will provide a comprehensive description of the equilibration processes occurring during the formation of interfacial layers in this type of system. It is expected that the results contained here may help to shed light on the complex physico-chemical behavior of these systems.

2. Materials and Methods

2.1. Chemicals

PDADMAC, with an average molecular weight in the 100–200 kDa range, was purchased as a 20 wt.% aqueous solution from Sigma-Aldrich (Saint Louis, MO, USA), and was used without further purification. SLES was supplied by Kao Chemical Europe S.L. (Barcelona, Spain) as an aqueous solution of surfactant concentration 70 wt.% and was purified by lyophilization followed by recrystallization of the obtained powder using acetone for HPLC (Acros Organics, Hampton, NH, USA) [50]. SLMT was synthetized and purified following the procedures described in a previous study [50]. Scheme 1 shows the molecular formula for PDADMAC and the two surfactants used in this work.

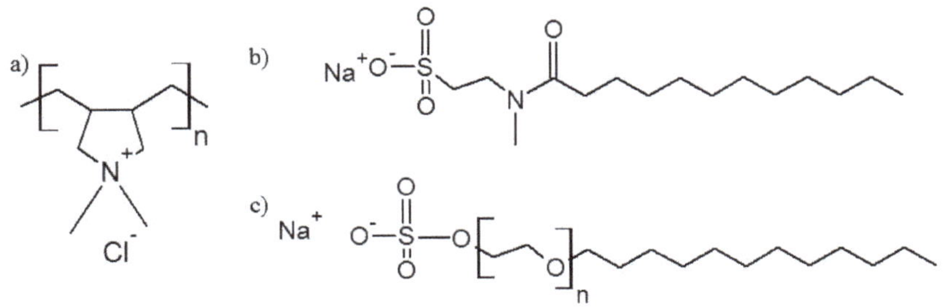

Scheme 1. Molecular formula of the three surfactants used in this work: PDADMAC (**a**), SLMT (**b**) and SLES (**c**).

Ultrapure deionized water used for cleaning and solution preparation was obtained using a multicartridge purification system AquaMAX^{TM}-Ultra 370 Series. (Young Lin, Anyang, Korea). This water presents a resistivity higher than 18 MΩ·cm, and a total organic content lower than 6 ppm. Glacial acetic acid and KCl (purity > 99.9%) purchased from Sigma-Aldrich were used to fix the pH and the ionic strength of solutions, respectively.

2.2. Preparation of Polyelectrolyte-Surfactant Solutions

The preparation of polyelectrolyte-surfactant solutions was performed following a procedure adapted from that proposed by Llamas et al. [53]. Firstly, the required amount of PDADMAC aqueous stock solution (concentration 20 wt.%) for obtaining a solution with polyelectrolyte concentration of 0.5 wt.% was weighted and poured into a flask. Then, KCl up to a final concentration of 40 mM was added into the flask. The last step involved the addition of the surfactant and the final dilution with an acetic acid solution of pH~5.6 to reach the final composition. The addition of surfactant was performed from stock aqueous solutions (pH~5.6) with a concentration one order of magnitude higher than that in the final solution. In this work, polyelectrolyte-surfactant solutions with surfactant concentration, c_s, in the range 10^{-6}–10 mM were studied. Once the solutions were prepared, these were mildly stirred (1000 rpm) for one hour using a magnetic stirrer to ensure the compositional homogenization of the solutions. Samples were left to age for 1 week prior to their use to ensure that no phase separation appeared in samples within the aging period [52].

2.3. Techniques

2.3.1. Turbidity Measurements

The turbidity of the solutions was evaluated from their transmittance at 400 nm, obtained using a UV-Visible spectrophotometer (HP-UV 8452, Hewlett Packard, Palo Alto, CA, USA). The turbidity of the samples was determined by the optical density at 400 nm ($OD_{400} = [100 - T(\%)]/100$, where T is the transmittance). It is worth mentioning that neither the polyelectrolyte nor the surfactant present any absorption band above 350 nm.

2.3.2. Binding Isotherm

The binding isotherm of the anionic surfactant to the polycation PDADMAC was determined by potentiometric titration using a surfactant selective electrode model 6.0507.120 from Metrohm (Herisau, Switzerland). The binding degree of surfactant β was estimated from the potentiometric measurements, as was proposed by Mezei and Meszaros [22]

$$\beta = \frac{c_s^{free}}{c_{monomer}} \tag{1}$$

where c_s^{free} and $c_{monomer}$ are the concentrations of free surfactant in solution and charged monomers of the polyelectrolyte chains, respectively. This method of determining the binding isotherm provides information about the amount of free surfactant remaining in the solution.

2.3.3. Surface Tension Measurements

Surface tension measurements as functions of the surfactant concentration (SLMT or SLES) for pure surfactant and polyelectrolyte-surfactant solutions were performed using different tensiometers. In all the cases, the adsorption was measured until the steady state conditions were reached. Special care was taken to limit the evaporation effects. Each value was obtained as an average of three independent measurements. All experiments were performed at 25.0 ± 0.1 °C. From the results of the experiments, it is possible to define the surface pressure as $\Pi(c_s) = \gamma_0 - \gamma(c_s)$, where γ_0 is the surface tension of the bare water–vapor interface and $\gamma(c_s)$ is the surface tension of the solution–vapor interface. Further details on surface tension experiments can be obtained from a previous study [23].

- Surface force tensiometers. Two different surface force tensiometers were used to measure the equilibrium surface tension: a surface force balance from Nima Technology (Coventry, UK), fitted with a disposable paper plate (Whatman CHR1 chromatography paper) as a contact probe; and a surface force tensiometer Krüss K10 (Hamburg, Germany), using a Pt Wilhelmy plate as a probe.
- Drop profile analysis tensiometer. A home-built drop profile analysis tensiometer in pendant drop configuration allowed determination of the surface tension of the water–vapor interface. This tensiometer enabled evaluation of the time dependence of the surface tension during the adsorption process, thus providing information related to the adsorption kinetics.

2.3.4. Dilational Rheology

A Nima 702 Langmuir balance from Nima Technology equipped with a surface force tensiometer was used to measure the response of the surface tension against sinusoidal changes in the surface area. Thus, it is possible to obtain information about the dilational viscoelatic moduli of the water–vapor interface $\varepsilon^* = \varepsilon' + i\varepsilon''$, with ε' and ε'' being the dilational elastic and viscous moduli, respectively, in the frequency range of 10^{-1}–10^{-2} Hz and at an area deformation amplitude $\Delta u = 0.1$, which was verified to be an appropriate value to ensure results within the linear regime of the layer response [52].

3. Results and Discussion

3.1. PDADMAC-Surfactants Assembly in Solution

The equilibrium condition implies that the chemical potential of all the species in both the bulk and at the interfaces are the same. Therefore, any physical understanding of the latter implies knowledge of the behavior of the different species in the bulk. Figure 1a shows the surfactant-binding isotherms deduced from electromotive force (EMF) measurements. Comparing the curves of EMF obtained for surfactants and PDADMAC-surfactant solutions, it is possible to obtain the binding isotherms for the corresponding surfactant to PDADMAC chains following the approach described by Mezei and Meszaros [50]. The results point out a high degree of binding over the whole range of studied compositions, providing an additional confirmation of the high efficiency of PDADMAC in binding anionic surfactants. Campbell et al. [38] found for PDADMAC-SDS solutions binding degrees of surfactant to PDADMAC close to 0.3 in the vicinity of the isoelectric point (surfactant concentration around 0.2 mM). The extrapolation of such results in similar conditions to those considered in this work, i.e., polymer concentration 50-fold the one used by Campbell et al. [38,52], and assuming that the binding is not significantly modified either for the surfactant structure or for the differences in the ionic strength, takes the binding degree at charge neutralization to a value <1%. This is just the situation found here, where binding isotherms evidence that the amount of free surfactant in solution remains below 10%, even for the highest surfactant concentrations. The low concentration of free surfactant in solution allows us to assume hereinafter that the bulk has a negligible free-surfactant concentration.

Figure 1b shows the dependence of the optical density of the samples on the surfactant concentration for the solutions of PDADMAC and the two surfactants. Similar qualitative concentration dependences of the optical density were found for both polyelectrolyte-surfactant systems. It may safely be expected that all of the studied compositions for PDADMAC-surfactant solutions fall in an equilibrium one-phase region, showing optically transparent solutions. This comes from the fact that the number of surfactant molecules available in solution is not high enough to neutralize the charge of all the monomers in the polyelectrolyte chains, thus leading to the formation of undercompensated cationic complexes in solution. Indeed, considering the high polymer concentration, simple calculations suggest the existence of around 36 monomers for each surfactant molecule for a surfactant concentration of approximately 1 mM. Therefore, assuming the complete binding of surfactant molecules to the polymer chains, around 35 monomers remain positively charged in the complexes, supporting the formation of transparent samples within the entire concentration range. However, contrary to what was expected for solutions with compositions far from the neutralization, the solutions formed by undercompensated complexes show an increase of the turbidity for the highest surfactant concentration. Therefore, for such concentrated solutions, the system should get close to the onset of where the two phase region occurs, even though no signature of charge neutralization was found from electrophoretic mobility measurements. This results from the mixing protocol used for solution preparation, which proceeds during the initial step by mixing a concentrated polymer solution with a concentrated surfactant solution. This precursor solution is them diluted up to the stated bulk composition. It may be expected that this methodology leads, due to the Marangoni stress created, to the formation of persistent kinetically-trapped aggregates that persist even upon dilution, leading to the appearance of a two-phase system far off the real neutralization point of the system [7,54]. These results contrast with those reported in other mixtures studied in the literature. In such systems, the increase on the optical density of the samples results from the formation of charge compensated complexes. The last is associated with the transition from a composition region, in which the charge of the complexes is governed by the excess of charged monomers to another region, in which the excess of bound surfactant to the polymer chain controls the charge of the formed complexes, i.e., a charge inversion transition [20,38,55]. The above results show that the production of kinetically-trapped aggregates during mixing can lead to turbid mixtures far from the real equilibrium phase separation [3]. Preliminary results have shown that the above discussed scenario changes significantly when the interaction of PDADMAC with betaine derived surfactants is considered. In such systems, even the polyelectrolyte-surfactant interactions occur through the negatively charged group in the terminal region of the polar head, the formation of kinetically-trapped aggregates is hindered, probably as a result of the electrostatic repulsion associated with the positively charged groups in the zwitterionic surfactant [56].

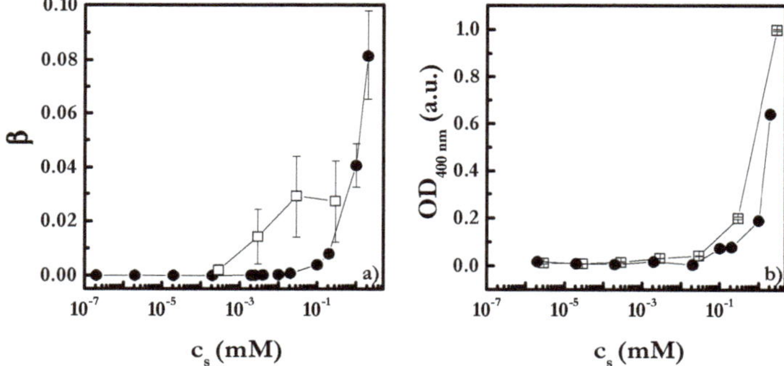

Figure 1. (a) Binding isotherms for surfactants on PDADMAC as a function of the initial concentration of surfactant in bulk. (b) Surfactant concentration dependences of the optical density of the solution, measured at 400 nm. Note: (□) = PDADMAC-SLMT; (●) = PDADMAC-SLES solutions. Lines are guides for the eyes. The results correspond to PDADMAC-surfactant mixtures containing a fixed PDADMAC concentration of 0.5 wt.%, and left to age for one week prior to measurement.

3.2. Equilibrium Adsorption at the Water–Vapor Interface

The evaluation of the surface pressure of solutions containing surface active compounds helps to understand the mechanisms involved in the equilibration of the water–vapor interface. Figure 2a shows the surface pressure dependences on the surfactant concentrations and on the PDADMAC concentration for the adsorption of the two surfactants and the polymer at the water–vapor interface (note that all solutions were prepared with the same pH and inert salt concentration as the polyelectrolyte-surfactant solutions). The results show that the surface activity of PDADMAC is negligible, at least up to concentrations that are 20-fold the one used in our work. This is in good agreement with the previous study by Noskov et al. [57] and with the negligible surface excess found for PDADMAC using neutron reflectrometry [38].

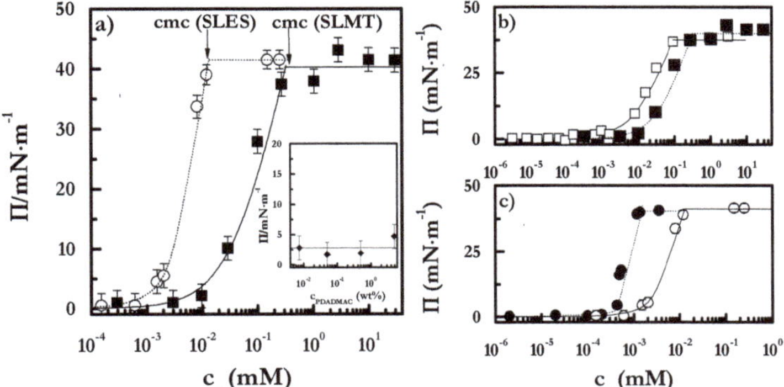

Figure 2. Results obtained using a drop profile analysis tensiometer: (**a**) Surface pressure dependence on surfactant concentration for the adsorption of pure SLES (○) and SLMT (■) at the water–vapor interface; cmc for both surfactants is marked. The inserted panel represents the surface pressure dependence on PDADMAC concentration for the adsorption of pure PDADMAC at the water–vapor interface. (**b**) Surface pressure dependence of SLMT concentration for pure SLMT (■) and PDADMAC–SLMT (□) solutions. (**c**) Surface pressure dependence of SLES concentration for pure SLES (○) and PDADMAC–SLES (●) solutions. The lines are guides for the eyes. The results for PDADMAC-surfactant mixtures correspond to mixtures containing a fixed PDADMAC concentration of 0.5 wt.%, and left to age for one week prior to measurement.

The adsorption behavior of SLMT and SLES is the expected for typical ionic surfactants. The Π increases with the bulk concentration up to the point that the surfactant concentration overcomes the threshold defined by the critical micellar concentration (cmc). Afterwards, Π remains constant with further increases of surfactant concentration. It is worth mentioning that the results obtained using different tensiometers (surface force tensiometer with Pt Wilhelmy as a probe plate and drop profile analysis tensiometer) agree within the combined error bars for the adsorption of both surfactants at the water–vapor interface. The surface pressure isotherms allow one to estimate the cmc of the pure surfactants, which showed values of around 10^{-2} and 10^{-1} mM for SLES and SLMT, respectively.

The comparison of the results obtained for the adsorption of pure surfactants at the water–vapor interface with those obtained for the adsorption of PDADMAC-surfactant solutions shows that for the lowest surfactant concentrations the surface pressure values are similar for pure surfactant and polyelectrolyte-surfactant solutions. This is the result of the low coverage of the interface (see Figure 2b,c). In such conditions, the surface excess is not high enough to produce any significant change in the surface free energy, and hence the Π values remain close to those of the bare water–vapor interface. The increase of the surfactant concentration leads to the increase of Π for both surfactant and polyelectrolyte-surfactant solutions. This increase starts for surfactant concentrations around one order of magnitude lower when polyelectrolyte-surfactant solutions are considered, which is a signature of the existence of a synergetic effect for the increase of the surface pressure as a result of the interaction in the solution of the polyelectrolyte and the surfactant. This is in agreement with previous results reported in the literature for several polyelectrolyte-surfactant systems [3,31,50,58]. The above-mentioned synergetic effects do not influence the adsorption behavior of solutions formed by PDADMAC and zwitterionic surfactants derived from the betaines, as was shown in preliminary results. This could be ascribed to the aforementioned differences in the aggregation process occurring in the bulk [56].

The study of the surface tension isotherms obtained for polymer-surfactant mixtures using different tensiometric techniques can help to understand the complexity of the interfacial behavior appearing when faced with these systems. Figure 3a,b shows that the surface tension isotherms obtained using different tensiometers reveal different features for PDADMAC-SLMT and PDADMAC-SLES solutions. PDADMAC-SLES solutions show similar surface pressure isotherms within the combined error bars, independent of the tensiometer used, and no evidences of the appearance of non-regular trends, either as surface tension peaks [38] or surface tension fluctuations [7], on the dependence of the surface pressure with the surfactant concentration were found. This contrasts with the results obtained for PDADMAC-SLMT solutions, in which the use of a surface force tensiometer with a Pt Wilhelmy plate as probe led to results that were significantly different to those obtained using the other tensiometers. The existence of such differences was previously reported in a study by Noskov et al. [31].

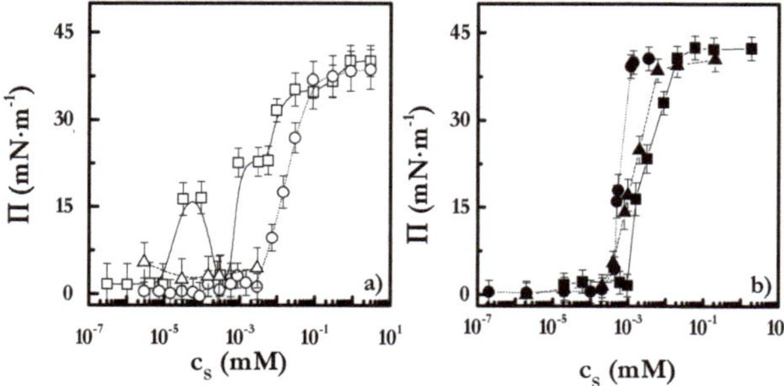

Figure 3. Surface pressure isotherms for solutions of PDADMAC with the two surfactants, obtained using different tensiometers. (**a**) Isotherms for PDADMAC-SLMT solutions. (**b**) Isotherms for PDADMAC-SLES solutions. Note: (□ and ■) Surface force tensiometer with Pt Wilhelmy plate as contact probe; (○ and ●) surface force tensiometer with paper Wilhelmy plate as contact probe; (Δ and ▲) drop profile analysis tensiometer. The lines are guides for the eyes. The results correspond to PDADMAC-surfactant mixtures containing a fixed PDADMAC concentration of 0.5 wt.% left to age for one week prior to measurement.

The differences found in the tensiometric behavior of PDADMAC-SLES and PDADMAC-SLMT solutions are correlated to differences in the equilibration mechanism of the interface. Assuming that the assembly of the polyelectrolyte-surfactant in solutions leads to the formation of kinetically trapped aggregates in both cases, this can evolve following different mechanisms upon adsorption at fluid interfaces. For PDADMAC-SLMT solutions, the appearance of surface tension fluctuations far from the phase separation region may be associated with the fact that upon adsorption at the water–vapor interface of the kinetically trapped aggregates can remain as isolated aggregates embedded at the interface. These do not dissociate spontaneously to form a kinetically trapped film at the interface. As a consequence, the trapped aggregates may adsorb onto the rough surface of the Pt Wilhelmy plate, changing its contact angle, which results in non-reliable surface tension values for the considered aggregates. This scenario is in agreement with the neutron reflectometry results obtained by Llamas et al. [50]. Their results showed a monotonic increase of the surface excess at the interface with the surfactant concentration, confirming that the surface tension fluctuations do not result from fluctuations of the interface composition. The behavior changes significantly when the adsorption of PDADMAC-SLES solutions is considered. In this case, the absence of surface tension fluctuation or significant differences in the results obtained using different tensiometers suggests the existence of dissociation and spreading of the kinetically trapped aggregates upon adsorption at the interface. Thus, the equilibration of the interface after the adsorption of the kinetically trapped aggregates occurs because of its dissociation, which is followed by the spreading of the complexes across the interface as a result of Marangoni flow associated with the lateral heterogeneity of the interface [38,42,50,59].

The differences in the adsorption mechanisms of PDADMAC-SLES and PDADMAC-SLMT complexes at the water–vapor interface may be explained on the bases of the molecular structures of the surfactant and the possibility to establish a cohesion interaction with the surrounding media. SLMT presents a hydrophobic tail formed by an alkyl chain, which tends to minimize the number of contact points with water, which favors the formed aggregates remaining as compact aggregates at the water–vapor interface upon adsorption. On the contrary, the presence of oxyethylene groups in SLES makes the dissociation and spreading of the complexes easier as a result of the possible formation of hydrogen bonds of the surfactant molecules with water. Surprisingly, studies on the adsorption of PDADMAC-SLES and PDADMAC-SLMT mixtures onto solid surfaces have evidenced a scenario compatible with that described for the adsorption at the fluid interfaces, where PDADMAC-SLES films present a topography reminiscent of the formation of extended complexes attached to the interface, whereas PDADMAC-SLMT films present a higher lateral heterogeneity [51,60]. Further confirmation of the discussed mechanisms may be obtained from the analysis of the adsorption kinetics at the water–vapor interface of the polyelectrolyte-surfactant solutions.

3.3. Adsorption Kinetics at the Water–Vapor Interface

The analysis of the adsorption kinetics of polymer-surfactants at the water–vapor interface is a powerful tool for deepening the understanding of the mechanistic aspects of the adsorption of complexes. This is done by studying the time evolution of the surface pressure (dynamic surface pressure) during the adsorption process. The adsorption kinetics have been measured using a drop shape analysis tensiometer. As expected, the adsorption of polymer-surfactant solutions at fluid interfaces is slower than that corresponding to pure surfactant [16,50]. Figure 4 shows the dynamics surface pressure obtained for the adsorption of PDADMAC-SLMT and PDADMAC-SLES solutions at the water–vapor interface.

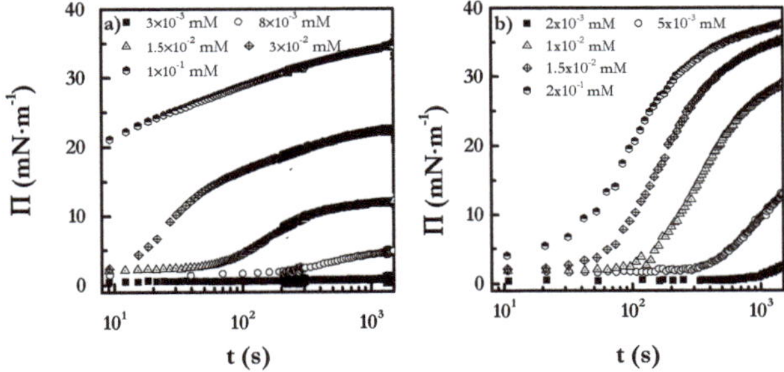

Figure 4. (a) Dynamic surface pressure for PDADMAC-SLMT solutions with different surfactant concentrations. (b) Dynamic surface pressure for PDADMAC-SLES solutions with different surfactant concentrations. The results correspond to PDADMAC-surfactant mixtures containing a fixed PDADMAC concentration of 0.5 wt.%, and left to age for one week prior to measurement.

The analysis of the adsorption kinetics show clearly that the increase of the surfactant concentration leads to the faster increase of the surface pressure, due to the higher hydrophobicity of the formed complexes. A more detailed analysis points out that whereas the adsorption of PDADMAC-SLMT is characterized by the monotonous increase of the surface pressure with time over the whole concentration range, the adsorption of PDADMAC-SLES presents an induction time that is reduced as the SLES concentration increases. Such differences are due to the differences in the processes involved in the equilibration of the interface.

The induction time in the adsorption of PDADMAC-SLES is explained considering that the equilibration of the interface proceeds following a two-step mechanism, as occurs for protein adsorption at fluid interfaces [61]. Firstly, polymer-surfactant complexes attach to the water–vapor interface as kinetically trapped aggregates until the surface excess overcomes a threshold value, after which point the adsorbed complexes undergo a dissociation and spreading process, which is responsible for the surface pressure increase [41,59]. It is worth mentioning that the decrease of the induction time with the increase of surfactant concentration results from the faster saturation of the interface, i.e., the shortening of the time needed to overcome the surface excess threshold, which leads to a prior surface pressure rise. The scenario found for PDADMAC-SLMT solutions is different to that described for PDADMAC-SLES, and the absence of the induction time is a signature of a difference in the equilibration mechanism of the interfacial layer. For PDADMAC-SLMT, the increase of the surface pressure is associated with the adsorption of isolated kinetically trapped aggregates that coalesce as the surfactant concentration increases. In this case, the adsorbed complexes remain compact without any significant dissociation. The above discussed results point out the existence of differences in the mechanisms for the equilibration of the interface of the polycation-anionic surfactant solution as result of the differences in the type of surfactant. The first one involves the dissociation and spreading of the pre-adsorbed kinetically trapped aggregates (PDADMAC-SLES), whereas the second one relies directly on the saturation of the interface with kinetically trapped aggregates. This proves that the adsorption of PDADMAC-SLMT leads to appreciable modifications of the surface pressure for surfactant concentrations one order of magnitude higher than PDADMAC-SLES as a result of the negligible effect of the isolated aggregates over the surface pressure of the bare water–vapor interface until their concentration is high enough. On the contrary, for PDADMAC-SLES, the dissociation and spreading of the aggregates enables the distribution of surface active material along the whole interface, and consequently the surface pressure starts to increase for lower surfactant concentrations as a result of the formation of interfacial layers in which complexes are extended along the interface.

3.4. Interfacial Dilational Rheology

The above discussion was devoted to the study of the adsorption at interfaces with fixed surface areas. However, from a technological point of view, the understanding of the response of the interface against external mechanical perturbations is essential because this provides important insights into the relaxation processes involved in the equilibration of interfacial layers [25,48,62,63]. The dependences of the dilational viscoelastic moduli (ε' represents the dilational elastic modulus and ε'' the viscous modulus) on the surfactant concentration and the deformation frequency provide complementary information for the better understanding of the complexity of the mechanism involved in the equilibration of the interfaces, helping to give a more detailed picture of the physical processes governing the formation of adsorption layers from polymer-surfactant solutions [64]. It must be stressed that for both PDADMAC-SLMT and PDADMAC-SLES solutions, the values of ε'' are negligible in relation to those of ε', with the ratio $\varepsilon''/\varepsilon'$ decreasing as the surfactant concentration increases. Hence, for the sake of simplicity only the behavior of ε' will be discussed. Figure 5 shows the concentration dependences of the elastic modulus for PDADMAC-SLMT and PDADMAC-SLES layers.

The results indicate that the dependence of ε' on the frequency is expected for the formation of layers at fluid interfaces, with ε' increasing with the deformation frequency. Furthermore, the concentration dependence of ε' is similar to that found for layers of surface active materials at fluid interfaces [46], with ε' increasing with the surfactant concentration from the value corresponding to the clean interface, reaching a maximum and then dropping down again to quasi-null values for the highest surfactant concentrations. A careful examination of the values obtained for the elasticity modulus for each system indicate that PDADMAC-SLES layers present values that are more than twice those obtained for PDADMAC-SLMT solutions independent of the considered frequency. This is again indicative of the different features of the interfacial layers. For PDADMAC-SLES layers, the spreading of material along the interface leads to the formation of extended complexes that can build a cross-linked network, increasing the elastic modulus of the interfacial layers. This cross-linking process is not possible when the interfacial layer is formed by compact kinetically trapped aggregates, as in PDADMAC-SLMT layers, leading to lower values of the elastic modulus of the interface.

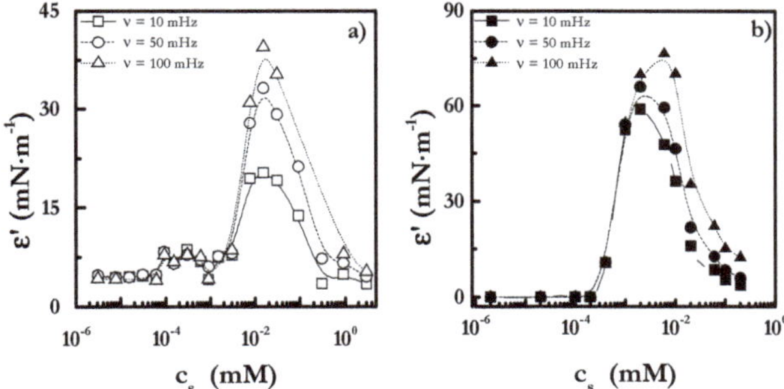

Figure 5. (a) Concentration dependences of the elastic modulus for PDADMAC-SLMT adsorption layers as were obtained from oscillatory barrier experiments performed at different frequencies. (b) Concentration dependences of the elastic modulus for PDADMAC-SLES adsorption layers, obtained from oscillatory barrier experiments performed at different frequencies. Note: ($\square$ and $\blacksquare$) $\nu = 0.01$ Hz; ($\bigcirc$ and $\bullet$) $\nu = 0.05$ Hz; ($\triangle$ and $\blacktriangle$) $\nu = 0.10$ Hz. The lines are guides for the eyes. For the sake of clarity, only results corresponding to some of the explored frequencies (ν) are shown, with the other frequencies presenting similar dependences. The results correspond to PDADMAC-surfactant mixtures containing a fixed PDADMAC concentration of 0.5 wt.%, and left to age for one week prior to measurement.

The frequency dependences of the elasticity modulus can be described in terms of the rheological model proposed by Ravera et al. [64,65] (see Figure 6a for an example). According to this model the frequency dependence of the viscoelastic modulus accounts for the initial adsorption of the polymer-surfactant complexes at the water–vapor interface as a diffusion-controlled process that is coupled to a second step associated with the internal reorganization of the adsorbed layers. Thus, taking into account the aforementioned framework, it is possible to describe the complex viscoelastic modulus with the following expression:

$$\varepsilon^* = \frac{1 + \xi + i\xi}{1 + 2\xi + 2\xi^2} \left[\varepsilon_0 + (\varepsilon_1 - \varepsilon_0) \frac{1 + i\lambda}{1 + \lambda^2} \right] \tag{2}$$

where $\xi = \sqrt{\nu_D / \nu}$, with ν_D and ν being the characteristic frequency of the diffusion exchange and the frequency of deformation, respectively, and $\lambda = \nu_1/\nu$, with ν_1 being the characteristic frequency of the extra relaxation process. Additionally, ε_0 and ε_1 represent the Gibbs elasticity and the high frequency limit elasticity within the frequency range considered, respectively. The validity of the discussed model, beyond confirming the complexity of the mechanisms involved in the equilibration of the interfacial layers formed by polyelectrolyte-surfactant solution, provides a description of the processes involved. It is expected that the equilibration of the interfacial occurs in the first stage by the diffusion-controlled adsorption of the kinetically trapped aggregates, and then such complexes undergo different reorganization processes depending on their nature. The existence of a two-step mechanism is in agreement with the picture proposed by Noskov et al. [45] for the equilibration of adsorption layers of PDADMAC-SDS at the water–vapor interface.

Figure 6b,c show the concentration dependences for the characteristic frequencies of the two dynamic processes appearing for the interfacial layers. As may be expected considering the different nature of the dynamic processes involved in the equilibration of the interfacial layer, ν_1, which is the frequency corresponding to the interfacial relaxation process, presents higher values than those associated with the diffusional transport, ν_D, for both PDADMAC-SLMT and PDADMAC-SLES solutions. This behavior can be explained by assuming that the interfacial relaxation process, involving the reorganization of materials at the interface, occurs only when a certain degree of material is adsorbed at the interface.

The results show that both ν_D and ν_1 increase in concentration for both studied systems. This increase can be explained in the case of ν_D as a result of the enhanced surface activity of the kinetically trapped aggregates, as the surfactant concentration increases due to their higher hydrophobicity. Furthermore, the values of ν_D are in a similar range for PDADMAC-SLMT and PDADMAC-SLES, which is in agreement with the similar origin of the process in both systems and the similarities of the complexes formed according to the above discussion. The slightly smaller values of ν_D found for PDADMAC-SLMT than for PDADMAC-SLES may result from different sizes of the complexes formed in the solution. The dependence of ν_1 is assumed to be because of the increase of surfactant in solution leading to an increase of the surface excess of complexes at the interface, which facilitates their reorganization within the interface. The higher values of ν_1 for PDADMAC-SLMT solution than PDADMAC-SLES solutions, at almost one order of magnitude, are ascribable again to the differences in the structure of the interfacial layers. Thus, the diffusion of extended complexes within the interface can occur across longer distances within the interface than that of compact aggregates, and consequently this process involves longer time scales.

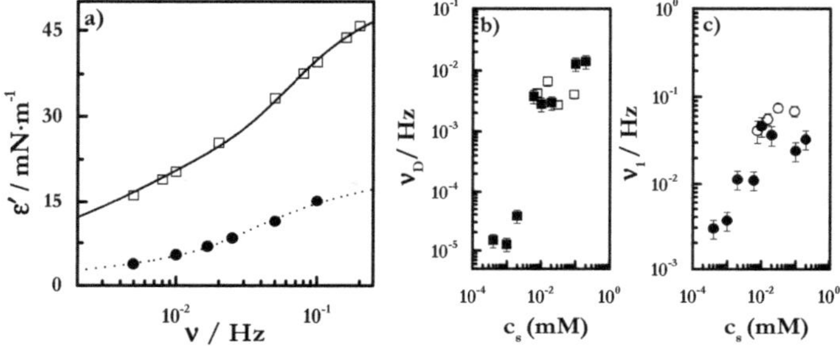

Figure 6. (a) Examples of frequency dependences of the elastic modulus for interfacial layers of PDADMAC-SLMT (□) and PDADMAC-SLES (●) and for solutions with surfactant concentration of 0.1 mM. Symbols represent the experimental data and the lines are the theoretical curves obtained from the analysis of the experimental results in term of the theoretical model described by Equation (2). (b) Concentration dependences of ν_D for PDADMAC-SLMT (□) and PDADMAC-SLES (■). (c) Concentration dependences of ν_1 for PDADMAC-SLMT (○) and PDADMAC-SLES (●). (b,c) Symbols represent the experimental data and the lines are guides for the eyes. The results correspond to PDADMAC-surfactant mixtures containing a fixed PDADMAC concentration of 0.5 wt.%, and left to age for one week prior to measurement.

4. Conclusions

The mechanisms involved in the equilibration of interfacial layers formed by adsorption of PDADMAC and two different anionic surfactants (SLMT and SLES) have been studied by surface tension (equilibrium and dynamics) and interfacial dilational rheology measurements. The combination of the interfacial characterization with studies on the association phenomena occurring in solution has evidenced that even the formation of kinetically trapped aggregates in the bulk occurs following similar patterns in both studied systems. These evolve following mechanisms depending of the specific chemical nature of the surfactant involved.

The equilibration of the interfacial layers formed by polyelectrolyte oppositely charged surfactants can be explained on the basis of a two-step mechanism. The first step is common to the different systems studied and is related with the diffusion-controlled incorporation of kinetically trapped aggregates at the water–vapor interfaces. Such aggregates can remain as compact aggregates at the interface, as in PDADMAC-SLMT solutions, or can undergo dissociation and spreading along the interface due to Marangoni flows, as in PDADMAC-SLES solutions. These different mechanisms result from differences in the hydrophobicity of the formed aggregates and the possibility to establish a cohesion interaction, such as a hydrogen bond, with the interface. On the basis of the obtained results, it can be concluded that there are no general laws governing the equilibration of the interfacial layers formed by the adsorption of polyelectrolyte-surfactant solutions at the fluid interface, with the process being primarily controlled by the specific nature of the chemical compounds involved and the interactions involved in the equilibration of the interface. This study contributes to the understanding of the fundamental basis describing the interfacial behavior of polyelectrolyte-surfactant solutions in conditions similar to that used in industrial application. Thus, the obtained result can help to exploit the interfacial behavior of these systems in technologically relevant conditions.

Author Contributions: Conceptualization, L.F.-P., A.A., and S.L.; Methodology, E.G., L.F.-P., A.A., and S.L.; Software, E.G.; Validation, E.G., F.O. and R.G.R.; Formal Analysis, E.G.; Investigation, E.G., L.F.P., A.A., S.L., F.O., and R.G.R.; Resources, R.G.R. and F.O.; Data Curation, E.G.; Writing—Original Draft Preparation, E.G.; Writing—Review and Editing, E.G., F.O. and R.G.R.; Visualization, E.G.; Supervision, E.G., F.O. and R.G.R.; Project Administration, R.G.R.; Funding Acquisition, R.G.R. and F.O.

Funding: This work was funded by MINECO under grant CTQ-2016-78895-R.

Acknowledgments: The CAI of spectroscopy from Universidad Complutense de Madrid is acknowledged for granting access to its facilities.

Conflicts of Interest: The authors declare no conflict of interest. The funders had no role in the design of the study; in the collection, analyses, or interpretation of data; in the writing of the manuscript, or in the decision to publish the results.

References

1. Llamas, S.; Guzmán, E.; Ortega, F.; Baghdadli, N.; Cazeneuve, C.; Rubio, R.G.; Luengo, G.S. Adsorption of polyelectrolytes and polyelectrolytes-surfactant mixtures at surfaces: A physico-chemical approach to a cosmetic challenge. *Adv. Colloid Interface Sci.* **2015**, *222*, 461–487. [CrossRef] [PubMed]

2. Goddard, E.D.; Ananthapadmanabhan, K.P. *Application of Polymer-Surfactant Systems*; Marcel Dekker, Inc.: New York, NY, USA, 1998.

3. Bain, C.D.; Claesson, P.M.; Langevin, D.; Meszaros, R.; Nylander, T.; Stubenrauch, C.; Titmuss, S.; von Klitzing, R. Complexes of surfactants with oppositely charged polymers at surfaces and in bulk. *Adv. Colloid Interface Sci.* **2010**, *155*, 32–49. [CrossRef] [PubMed]

4. Khan, N.; Brettmann, B. Intermolecular interactions in polyelectrolyte and surfactant complexes in solution. *Polymers* **2019**, *11*, 51. [CrossRef] [PubMed]

5. Gradzielski, M.; Hoffmann, I. Polyelectrolyte-surfactant complexes (PESCs) composed of oppositely charged components. *Curr. Opin. Colloid Interface Sci.* **2018**, *35*, 124–141. [CrossRef]

6. Guzmán, E.; Llamas, S.; Maestro, A.; Fernández-Peña, L.; Akanno, A.; Miller, R.; Ortega, F.; Rubio, R.G. Polymer-surfactant systems in bulk and at fluid interfaces. *Adv. Colloid Interface Sci.* **2016**, *233*, 38–64. [CrossRef] [PubMed]

7. Varga, I.; Campbell, R.A. General physical description of the behavior of oppositely charged polyelectrolyte/surfactant mixtures at the air/water interface. *Langmuir* **2017**, *33*, 5915–5924. [CrossRef] [PubMed]

8. Nylander, T.; Samoshina, Y.; Lindman, B. Formation of polyelectrolyte-surfactant complexes on surfaces. *Adv. Colloid Interface Sci.* **2006**, *123–126*, 105–123. [CrossRef]

9. Ferreira, G.A.; Loh, W. Liquid crystalline nanoparticles formed by oppositely charged surfactant-polyelectrolyte complexes. *Curr. Opin. Colloid Interface Sci.* **2017**, *32*, 11–22. [CrossRef]

10. Piculell, L.; Lindman, B. Association and segregation in aqueos polymer/polymer, polymer/surfactant, and surfactant/surfactant mixtures: Similarities and differences. *Adv. Colloid Interface Sci.* **1992**, *41*, 149–178. [CrossRef]

11. Liu, J.Y.; Wang, J.G.; Li, N.; Zhao, H.; Zhou, H.J.; Sun, P.C.; Chen, T.H. Polyelectrolyte-surfactant complex as a template for the synthesis of zeolites with intracrystalline mesopores. *Langmuir* **2012**, *28*, 8600–8607. [CrossRef]

12. Miyake, M. Recent progress of the characterization of oppositely charged polymer/surfactant complex in dilution deposition system. *Adv. Colloid Interface Sci.* **2017**, *239*, 146–157. [CrossRef] [PubMed]

13. Szczepanowicz, K.; Bazylińska, U.; Pietkiewicz, J.; Szyk-Warszyńska, L.; Wilk, K.A.; Warszyński, P. Biocompatible long-sustained release oil-core polyelectrolyte nanocarriers: From controlling physical state and stability to biological impact. *Adv.Colloid Interface Sci.* **2015**, *222*, 678–691. [CrossRef] [PubMed]

14. Picullel, L. Understanding and exploiting the phase behavior of oppositely charged polymer and surfactant in water. *Langmuir* **2013**, *29*, 10313–10329. [CrossRef] [PubMed]

15. Goddard, E.D. Polymer/surfactant interaction: Interfacial aspects. *J. Colloid Interface Sci.* **2002**, *256*, 228–235. [CrossRef]

16. Llamas, S.; Guzmán, E.; Akanno, A.; Fernández-Peña, L.; Ortega, F.; Campbell, R.A.; Miller, R.; Rubio, R.G. Study of the liquid/vapor interfacial properties of concentrated polyelectrolyte-surfactant mixtures using surface tensiometry and neutron reflectometry: Equilibrium, adsorption kinetics, and dilational rheology. *J. Phys. Chem. C* **2018**, *122*, 4419–4427. [CrossRef]

17. Campbell, R.A.; Arteta, M.Y.; Angus-Smyth, A.; Nylander, T.; Varga, I. Effects of bulk colloidal stability on adsorption layers of poly(diallyldimethylammonium chloride)/sodium dodecyl sulfate at the air/water interface studied by neutron reflectometry. *J. Phys. Chem. B* **2011**, *115*, 15202–15213. [CrossRef] [PubMed]

18. Campbell, R.A.; Arteta, M.Y.; Angus-Smyth, A.; Nylander, T.; Varga, I. Multilayers at interfaces of an oppositely charged polyelectrolyte/surfactant system resulting from the transport of bulk aggregates under gravity. *J. Phys. Chem B* **2012**, *116*, 7981–7990. [CrossRef] [PubMed]

19. Campbell, R.A.; Arteta, M.Y.; Angus-Smyth, A.; Nylander, T.; Noskov, B.A.; Varga, I. Direct impact of non-equilibrium aggregates on the structure and morphology of pdadmac/SDS layers at the air/water interface. *Langmuir* **2014**, *30*, 8664–8774. [CrossRef] [PubMed]

20. Mészáros, R.; Thompson, L.; Bos, M.; Varga, I.; Gilányi, T. Interaction of sodium dodecyl sulfate with polyethyleneiminie: surfactant-induced polymer solution colloid dispersion transition. *Langmuir* **2003**, *19*, 609–615. [CrossRef]

21. Mezei, A.; Pojják, K.; Mészaros, R. Nonequilibrium features of the association between poly(vinylamine) and sodium dodecyl sulfate: The validity of the colloid dispersion concept. *J. Phys. Chem B* **2008**, *112*, 9693–9699. [CrossRef] [PubMed]

22. Pojják, K.; Bertalanits, E.; Mészáros, R. Effect of salt on the equilibrium and nonequilibrium features of polyelectrolyte/surfactant association. *Langmuir* **2011**, *27*, 9139–9147. [CrossRef] [PubMed]

23. Bodnár, K.; Fegyver, E.; Nagy, M.; Mészáros, R. Impact of polyelectrolyte chemistry on the thermodynamic stability of oppositely charged macromolecules/surfactant mixtures. *Langmuir* **2016**, *32*, 1259–1268. [CrossRef] [PubMed]

24. Goddard, E.D.; Hannan, R.B. Cationic polymer/anionic surfactant interactions. *J. Colloid Interface Sci.* **1976**, *55*, 73–79. [CrossRef]

25. Bergeron, V.; Langevin, D.; Asnacios, A. Thin-film forces in foam films containing anionic polyelectrolyte and charged surfactants. *Langmuir* **1996**, *12*, 1550–1556. [CrossRef]

26. Bhattacharyya, A.; Monroy, F.; Langevin, D.; Argillier, J.-F. Surface rheology and foam stability of mixed surfactant-polyelectrolyte solutions. *Langmuir* **2000**, *16*, 8727–8732. [CrossRef]

27. Stubenrauch, C.; Albouy, P.-A.; von Klitzing, R.; Langevin, D. Polymer/surfactant complexes at the water/air interface: A surface tension and x-ray reflectivity study. *Langmuir* **2000**, *16*, 3206–3213. [CrossRef]

28. Braun, L.; Uhlig, M.; von Klitzing, R.; Campbell, R.A. Polymers and surfactants at fluid interfaces studied with specular neutron reflectometry. *Adv. Colloid Interface Sci.* **2017**, *247*, 130–148. [CrossRef]

29. Lu, J.R.; Thomas, R.K.; Penfold, J. Surfactant layers at the air/water interface: Structure and composition. *Adv. Colloid Interface Sci.* **2000**, *84*, 143–304. [CrossRef]

30. Narayanan, T.; Wacklin, H.; Konovalov, O.; Lund, R. Recent applications of synchrotron radiation and neutrons in the study of soft matter. *Crystallography Rev.* **2017**, *23*, 160–226. [CrossRef]

31. Staples, E.; Tucker, I.; Penfold, J.; Warren, N.; Thomas, R.K.; Taylor, D.J.F. Organization of polymer−surfactant mixtures at the air−water interface: sodium dodecyl sulfate and poly(dimethyldiallylammonium chloride). *Langmuir* **2002**, *18*, 5147–5153. [CrossRef]

32. Penfold, J.; Tucker, I.; Thomas, R.K.; Zhang, J. Adsorption of polyelectrolyte/surfactant mixtures at the air−solution interface:poly(ethyleneimine)/sodium dodecyl sulfate. *Langmuir* **2005**, *21*, 10061–10073. [CrossRef] [PubMed]

33. Penfold, J.; Thomas, R.K.; Taylor, D.J.F. Polyelectrolyte/surfactant mixtures at the air–solution interface. *Curr. Opin. Colloid Interface Sci.* **2006**, *11*, 337–344. [CrossRef]
34. Penfold, J.; Tucker, I.; Thomas, R.K.; Taylor, D.J.F.; Zhang, X.L.; Bell, C.; Breward, C.; Howell, P. The interaction between sodium alkyl sulfate surfactants and the oppositely charged polyelectrolyte, polyDMDAAC, at the air-water interface: The role of alkyl chain length and electrolyte and comparison with theoretical predictions. *Langmuir* **2007**, *23*, 3128–3136. [CrossRef] [PubMed]
35. Thomas, R.K.; Penfold, J. Thermodynamics of the air-water interface of mixtures of surfactants with polyelectrolytes, oligoelectrolyte, and multivalent metal electrolytes. *J. Phys. Chem B* **2018**, *122*, 12411–12427. [CrossRef] [PubMed]
36. Bell, C.G.; Breward, C.J.W.; Howell, P.D.; Penfold, J.; Thomas, R.K. A theoretical analysis of the surface tension profiles of strongly interacting polymer–surfactant systems. *J. Colloid Interface Sci.* **2010**, *350*, 486–493. [CrossRef] [PubMed]
37. Bahramian, A.; Thomas, R.K.; Penfold, J. The adsorption behavior of ionic surfactants and their mixtures with nonionic polymers and with polyelectrolytes of opposite charge at the air–water interface. *J. Phys. Chem. B* **2014**, *118*, 2769–2783. [CrossRef] [PubMed]
38. Campbell, R.A.; Angus-Smyth, A.; Yanez-Arteta, M.; Tonigold, K.; Nylander, T.; Varga, I. New perspective on the cliff edge peak in the surface tension of oppositely charged polyelectrolyte/surfactant mixtures. *J. Phys. Chem. Lett.* **2010**, *1*, 3021–3026. [CrossRef]
39. Ábraham, A.; Campbell, R.A.; Varga, I. New method to predict the surface tension of complex synthetic and biological polyelectrolyte/surfactant mixtures. *Langmuir* **2013**, *29*, 11554–11559. [CrossRef] [PubMed]
40. Angus-Smyth, A.; Bain, C.D.; Varga, I.; Campbell, R.A. Effects of bulk aggregation on PEI–SDS monolayers at the dynamic air–liquid interface: Depletion due to precipitation versus enrichment by a convection/spreading mechanism. *Soft Matter* **2013**, *9*, 6103–6117. [CrossRef]
41. Campbell, R.A.; Tummino, A.; Noskov, B.A.; Varga, I. Polyelectrolyte/surfactant films spread from neutral aggregates. *Soft Matter* **2016**, *12*, 5304–5312. [CrossRef] [PubMed]
42. Noskov, B.A.; Loglio, G.; Miller, R. Dilational surface visco-elasticity of polyelectrolyte/surfactant solutions: Formation of heterogeneous adsorption layers. *Adv. Colloid Interface Sci.* **2011**, *168*, 179–197. [CrossRef] [PubMed]
43. Lyadinskaya, V.V.; Bykov, A.G.; Campbell, R.A.; Varga, I.; Lin, S.Y.; Loglio, G.; Miller, R.; Noskov, B.A. Dynamic surface elasticity of mixed poly(diallyldimethylammoniumchloride)/sodium dodecyl sulfate/NaCl solutions. *Colloids Surf. A* **2014**, *460*, 3–10. [CrossRef]
44. Monteux, C.; Fuller, G.G.; Bergeron, V. Shear and dilational surface rheology of oppositely charged polyelectrolyte/surfactant microgels adsorbed at the air-water interface. Influence on foam stability. *J. Phys. Chem. B* **2004**, *108*, 16473–16482. [CrossRef]
45. Noskov, B.A.; Grigoriev, D.O.; Lin, S.Y.; Loglio, G.; Miller, R. Dynamic surface properties of polyelectrolyte/surfactant adsorption films at the air/water interface: Poly(diallyldimethylammonium chloride) and sodium dodecylsulfate. *Langmuir* **2007**, *23*, 9641–9651. [CrossRef] [PubMed]
46. Noskov, B.A. Dilational surface rheology of polymer and polymer/surfactant solutions. *Curr. Opin. Colloids Interface Sci.* **2010**, *15*, 229–236. [CrossRef]
47. Fauser, H.; von Klitzing, R.; Campbell, R.A. Surface adsorption of oppositely charged C14TAB-PAMPS mixtures at the air/water interface and the impact on foam film stability. *J. Phys. Chem. B* **2015**, *119*, 348–358. [CrossRef]
48. Fuller, G.G.; Vermant, J. Complex fluid-fluid interfaces: Rheology and structure. *Annu. Rev. Chem. Biomol. Eng.* **2012**, *3*, 519–543. [CrossRef]
49. Regismond, S.T.A.; Winnik, F.M.; Goddard, E.D. Surface viscoelasticity in mixed polycation anionic surfactant systems studied by a simple test. *Colloids Surf. A* **1996**, *119*, 221–228. [CrossRef]
50. Llamas, S.; Fernández-Peña, L.; Akanno, A.; Guzmán, E.; Ortega, V.; Ortega, F.; Csaky, A.G.; Campbell, R.A.; Rubio, R.G. Towards understanding the behavior of polyelectrolyte—Surfactant mixtures at the water/vapor interface closer to technologically-relevant conditions. *Phys. Chem. Chem. Phys.* **2018**, *20*, 1395–1407. [CrossRef]
51. Llamas, S.; Guzmán, E.; Baghdadli, N.; Ortega, F.; Cazeneuve, C.; Rubio, R.G.; Luengo, G.S. Adsorption of poly(diallyldimethylammonium chloride)-sodium methyl-cocoyl-taurate complexes onto solid surfaces. *Colloids Surf. A* **2016**, *505*, 150–157. [CrossRef]

52. Mendoza, A.J.; Guzmán, E.; Martínez-Pedrero, F.; Ritacco, H.; Rubio, R.G.; Ortega, F.; Starov, V.M.; Miller, R. Particle laden fluid interfaces: Dynamics and interfacial rheology. *Adv.Colloid Interface Sci.* **2014**, *206*, 303–319. [CrossRef] [PubMed]

53. Goddard, E.D.; Gruber, J.V. *Principles of Polymer Science and Technology in Cosmetics and Personal Care*; Marcel Dekker, Inc.: Basel, Switzerland, 1999.

54. Mezei, A.; Mezaros, R. Novel method for the estimation of the binding isotherms of ionic surfactants on oppositely charged polyelectrolytes. *Langmuir* **2006**, *22*, 7148–7151. [CrossRef] [PubMed]

55. Naderi, A.; Claesson, P.M.; Bergström, M.; Dedinaite, A. Trapped non-equilibrium states in aqueous solutions of oppositely charged polyelectrolytes and surfactants: Effects of mixing protocol and salt concentration. *Colloids Surf. A* **2005**, *253*, 83–93. [CrossRef]

56. Akanno, A. *Bulk and Surface Properties of Polyelectrolyte Surfactant Mixtures*; Universidad Complutense de Madrid: Madrid, Spain, 2018.

57. Mezei, A.; Mészáros, R.; Varga, I.; Gilanyi, T. Effect of mixing on the formation of complexes of hyperbranched cationic polyelectrolytes and anionic surfactants. *Langmuir* **2007**, *23*, 4237–4247. [CrossRef] [PubMed]

58. Noskov, B.A.; Bilibin, A.Y.; Lezov, A.V.; Loglio, G.; Filippov, S.K.; Zorin, I.M.; Miller, R. Dynamic surface elasticity of polyelectrolyte solutions. *Colloids Surf. A* **2007**, *298*, 115–122. [CrossRef]

59. Tummino, A.; Toscano, J.; Sebastiani, F.; Noskov, B.A.; Varga, I.; Campbell, R.A. Effects of aggregate charge and subphase ionic strength on the properties of spread polyelectrolyte/surfactant films at the air/water interface under static and dynamic conditions. *Langmuir* **2018**, *34*, 2312–2323. [CrossRef]

60. Llamas, S. *Estudio de Interfases de Interés en Cosmética*; Universidad Complutense de Madrid: Madrid, Spain, 2014.

61. Erickson, J.S.; Sundaram, S.; Stebe, K.J. Evidence that the induction time in the surface pressure evolution of lysozyme solutions is caused by a surface phase transition. *Langmuir* **2000**, *16*, 5072–5078. [CrossRef]

62. Schramm, L.L. *Emulsions, Foams, Suspensions, and Aerosols*; Wiley-VCH Verlag GmbH & Co.: Weinheim, Germany, 2014.

63. Langevin, D. Aqueous foams: A field of investigation at the frontier between chemistry and physics. *ChemPhysChem* **2008**, *9*, 510–522. [CrossRef]

64. Liggieri, L.; Santini, E.; Guzmán, E.; Maestro, A.; Ravera, F. Wide-frequency dilational rheology investigation of mixed silica nanoparticle—CTAB interfacial layers. *Soft Matter* **2011**, *7*, 6699–7709. [CrossRef]

65. Ravera, F.; Ferrari, M.; Santini, E.; Liggieri, L. Influence of surface processes on the dilational visco-elasticity of surfactant solutions. *Adv. Colloid Interface Sci.* **2005**, *117*, 75–100. [CrossRef]

Article

Studies on Synthesis and Characterization of Aqueous Hybrid Silicone-Acrylic and Acrylic-Silicone Dispersions and Coatings. Part I

Janusz Kozakiewicz [1,*], Joanna Trzaskowska [1], Wojciech Domanowski [1], Anna Kieplin [2], Izabela Ofat-Kawalec [1], Jarosław Przybylski [1], Monika Woźniak [2], Dariusz Witwicki [2,†] and Krystyna Sylwestrzak [1]

[1] Department of Polymer Technology and Processing, Industrial Chemistry Research Institute, Rydygiera 8, 01-793 Warsaw, Poland; joanna.trzaskowska@ichp.pl (J.T.); wojciech.domanowski@ichp.pl (W.D.); izabela.ofat-kawalec@ichp.pl (I.O.-K.); jaroslaw.przybylski@ichp.pl (J.P.); krystyna.sylwestrzak@ichp.pl (K.S.)

[2] D&R Dispersions and Resins Sp. z o. o., Duninowska 9, 87-800 Włocławek, Poland; anna.kieplin@d-resins.com (A.K.); monika.wozniak@d-resins.com (M.W.)

[*] Correspondence: janusz.kozakiewicz@ichp.pl; Tel.: +48-500-433-297

[†] The author passed away (1963–2018).

Received: 15 November 2018; Accepted: 23 December 2018; Published: 3 January 2019

Abstract: The objective of the study was to investigate the effect of the method of synthesis on properties of aqueous hybrid silicone-acrylic (SIL-ACR) and acrylic-silicone (ACR-SIL) dispersions. SIL-ACR dispersions were obtained by emulsion polymerization of mixtures of acrylic and styrene monomers (butyl acrylate, styrene, acrylic acid and methacrylamide) of two different compositions in aqueous dispersions of silicone resins synthesized from mixtures of silicone monomers (octamethylcyclotetrasiloxane, vinyltriethoxysilane and methyltriethoxysilane) of two different compositions. ACR-SIL dispersions were obtained by emulsion polymerization of mixtures of the same silicone monomers in aqueous dispersions of acrylic/styrene copolymers synthesized from the same mixtures of acrylic and styrene monomers, so the compositions of ACR and SIL parts in corresponding ACR-SIL and SIL-ACR hybrid dispersions were the same. Examination of the properties of hybrid dispersions (particle size, particle structure, minimum film forming temperature, T_g of dispersion solids) as well as of corresponding coatings (contact angle, water resistance, water vapour permeability, impact resistance, elasticity) and films (tensile strength, elongation at break, % swell in toluene), revealed that they depended on the method of dispersion synthesis that led to different dispersion particle structures and on composition of ACR and SIL part. Generally, coatings produced from hybrid dispersions showed much better properties than coatings made from starting acrylic/styrene copolymer dispersions.

Keywords: aqueous polymer dispersions; silicone-acrylic; acrylic-silicone; hybrid particle structure; coatings

---

1. Introduction

Aqueous polymer dispersions are currently produced in quantities exceeding globally 20 million tons per annum [1] and are commonly used, inter alia, as binders for organic coatings, especially for aqueous dispersion-based architectural paints. The reason for a great increase in research and business interest in that specific group of products is not only the fact that they are environmentally friendly, but also the possibility of tailoring the composition and structure of dispersion particle in order to achieve desired performance characteristics of the final coating. If the particles have a hybrid (It is worth to note that generally a "hybrid material" is a material that is composed of at least two components mixed at molecular scale [2] and although this term is normally used for polymer-inorganic structure composites [3], it can be also applied to polymer-polymer systems) structure (i.e., are composed of at least two different polymers) and their diameter is less than 100 nm they may be called "dispersion nanoparticles with hybrid structure" within which the occurrence of specific interactions between these polymers optionally leading to synergistic effect may be expected. Then, due to a synergistic effect, new and sometimes quite unexpected features of coatings or films made using such hybrid dispersions as binders may be found—see Figure 1.

(a)   (b)

Figure 1. Differences between mixture of two aqueous dispersions of different polymers (polymer 1—blue color and polymer 2—red color) and aqueous dispersion with hybrid particle structure composed of the same two different polymers. In the mixture of two dispersions (**a**) synergistic effect is much less probable than in dispersion with hybrid particle structure (**b**) [4].

Although some authors reported that certain specific properties of coatings could be improved by using blends of dispersions of different polymers (e.g., dirt resistance could be enhanced this way [5]), clear advantages of application of dispersions with hybrid particle structure as coating binders have been confirmed in the literature, e.g., [1,4,6] and descriptions of methods applied for synthesis of such dispersions can be found in books and review papers, e.g., [7–10]. The particle morphology that is most frequently referred to in the research articles is a "core-shell" structure, but other morphologies, like core-double shell, gradient, eye-ball-like, raspberry-like, fruit cake or embedded sphere can also be obtained [11,12]. It has been proved [8] that not only the hybrid dispersion particle size and chemical composition, but also its morphology can significantly affect the properties of both dispersions and coatings produced from such dispersions. Therefore, investigation of the hybrid dispersion systems is of great interest to many researchers.

According to [7], the following general approaches can be applied to synthesis of hybrid aqueous dispersions in the emulsion polymerization process:

- A process where monomer X is polymerized in aqueous dispersion of polymer Y or monomer Y is polymerized in aqueous dispersion of polymer X;
- A process where monomer X is added to aqueous dispersion of polymer Y or monomer Y is added to aqueous dispersion of polymer X and left for some time in order to achieve swelling of dispersion particles with the monomer, and only then is polymerization conducted;
- A process where a mixture of monomers X and Y is placed in the reactor before start of polymerization or is added dropwise during the polymerization. However, in this case formation of particles with hybrid structure would be possible only if the corresponding homopolymers are not compatible or either reactivities of monomers or their polymerization mechanisms differ significantly.

As is clear from the literature, e.g., [4,13] the aforementioned methods of hybrid aqueous dispersion synthesis can be successfully applied to obtain different dispersion particle morphologies, depending on the selection of the starting materials (polymers, monomers, surfactants, initiators etc.). It can be expected that if methods (1) or (2) are applied, the "core-shell" morphology will be the most probable one supposing that certain conditions are fulfilled: "core" polymer should be more hydrophobic than "shell" polymer and formation of separate particles of polymer X in the course of polymerization in dispersion of polymer Y resulting from homolytic nucleation is retarded, e.g., by diminishing the monomer and surfactant concentration in the reaction mixture. The mechanism of hybrid particle formation in the emulsion polymerization process has been described in detail, e.g., in [4,13–15] and factors that determine creation of specific particle morphology have been identified. A review of fundamental theoretical aspects of the formation of dispersion particles with a hybrid structure has been published [16].

One of the hybrid aqueous dispersion systems that is most interesting from the point of view of practical application, especially as coating binders, is dispersion with particles containing organic polymer (usually polyacrylate) and silicone. This is because silicone-containing polymer systems allow for achieving specific features of coating surface like e.g., water repellency without affecting its general performance [17]. A comprehensive review of synthesis and characterization of such hybrid silicone-acrylic dispersions as well as of coatings and films or powders produced from them has been published in 2015 [11]. It has been proved in a number of research papers both referred to in that review paper and published later that the unique properties of coatings like e.g., high surface hydrophobicity and water resistance combined with enhanced water vapour permeability and good mechanical properties can be achieved by applying methods (1) to (3) described above to synthesize hybrid dispersions containing silicones where monomer X is acrylic monomer and monomer Y is silicone monomer—see e.g., [18–21] for method (1), [22–24] for method (2) or [24–26] for method (3). If fluoroacrylic monomer was used as monomer X [27–30], the surface hydrophobicity of coatings could be improved even more. It was also proved in our earlier studies [31–33] that the application of method (1) to emulsion polymerization of methyl methacrylate in aqueous silicone resin dispersions led to stable "core-shell" silicone-poly(methyl methacrylate) hybrid dispersions which, after drying, produced corresponding "nanopowders" that were later used as very effective impact modifiers for powder coatings.

In the present study we investigated the effect of the approach to synthesis (method (1) or method (2) as defined above) on the properties of hybrid aqueous dispersions and corresponding coatings. Two different silicone resin dispersions and two different acrylic/styrene copolymer dispersions were used as starting media in which emulsion polymerization of acrylic and styrene monomers or silicone monomers respectively was conducted. The mass ratio equal to 1/3 of silicone part (SIL 1 or SIL 2) to acrylic/styrene part (ACR A or ACR B) in the synthesis was selected based on the assumption supported by the literature [11] that this proportion would be sufficient to observe the influence of the presence of silicone in the dispersion particle on the properties of hybrid dispersions and coatings. Further studies on the effect of SIL/ACR ratio on the properties of hybrid dispersions and coatings are ongoing and will be published soon.

2. Materials and Methods

2.1. Starting Materials

Octamethylcyclotetrasiloxane (D4) was obtained from Momentive (Waterford, NY, USA). Other silicone monomers: vinyltriethoxysilane (VTES) and methyltriethoxysilane (MTES)) were obtained from Evonik (Essen, Germany) under trade names Dynasylan® VTEO and Dynasylan® MTES. Surfactants dodecylbenzenesulphonic acid (DBSA) and Rokanol T18 (nonionic surfactant based on ethoxylated C16–C18 alcohols) were obtained from PCC Exol (Brzeg Dolny, Poland). Emulgator E30 (anionic surfactant based on C15 alkylsulfonate) was obtained from LeunaTenside GmbH (Leuna, Germany). Other standard ingredients used in the synthesis of dispersions (sodium acetate, sodium hydrocarbonate, potassium persulphate and aqueous ammonia solution) were obtained from Standard Lublin (Poland) as pure reagents. Biocide used to protect dispersions from infestation was Acticide MBS obtained from THOR (Wincham, UK). Starting acrylic/styrene copolymer dispersions (ACR A and ACR B) characterized by different T_{gs} were supplied by Dispersions & Resins (D&R, Włocławek, Poland). Monomers applied in synthesis of ACR A and ACR B dispersions by D&R were butyl acrylate (BA) obtained from ECEM, Arkema, Indianapolis, IN, USA, styrene (ST) obtained from KH Chemicals, Helm, Zwijndrecht, The Netherlands, acrylic acid (AA) obtained from Prochema, BASF, Wien, Austria, and methacrylamide (MA) obtained from ECEM, Arkema. Acrylic and styrene monomers were used as received as mixtures designated as A and B with compositions corresponding to compositions of monomers applied to synthesize dispersions ACR A and ACR B, respectively. Exact compositions could not be revealed due to commercial secret, but were appropriately designed to get T_g of dispersion solids at a level of ca. +15 °C (ACR A) and of ca. +30 °C (ACR B). For structures of acrylic monomers-see Figure 2.

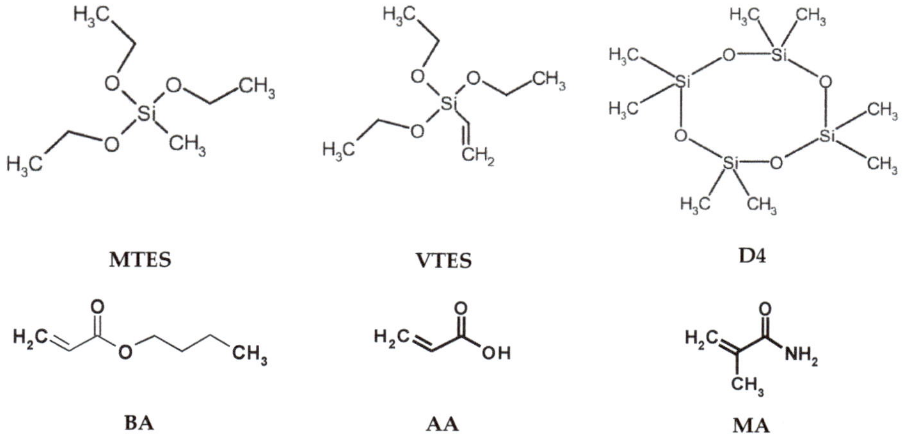

Figure 2. Structures of silicone monomers used in synthesis of silicone resin dispersions SIL 1 and SIL 2 and acrylic/styrene polymer dispersions ACR A and ACR B.

ACR A and ACR B dispersions were not neutralized after polymerization had been completed in order to ensure the low pH value (ca. 3) that was needed to conduct polymerization of silicone monomers in the process of synthesis of hybrid acrylic-silicone dispersions.

2.2. Synthesis of Silicone Resin Dispersions and Hybrid Silicone-Acrylic and Acrylic-Silicone Dispersions

Silicone resin dispersions (SIL 1 and SIL 2) were synthesized according to the procedure described in [31], although different functional silanes were used along with D4 as silicone monomers—see Figure 2 for the structures of these silicone monomers.

The compositions (wt.%) of mixtures of silicone monomers used in synthesis of SIL 1 and SIL 2 were as follows:

- Mixture designated as 1: D4—84.0%, MTES—9.5%, VTES—6.5%
- Mixture designated as 2: D4—88.0%, VTES—12%

DBSA was used as surfactant and D4 ring-opening catalyst. The reaction that proceeded in the process of SIL 1 and SIL 2 synthesis was simultaneous hydrolysis of trifunctional ethoxysilanes and breaking of Si-O bond in D4 leading to the formation of partially crosslinked poly(dimethylsiloxane) containing unsaturated bonds originating from VTES (see Figure 3).

After distillation of ethanol under vacuum no free VTES or MTES were detected by gas chromatography (GC) in the resulting SIL dispersions, although small amounts of D4 (ca. 0.8%) and ethanol (ca. 0.2%) were still present.

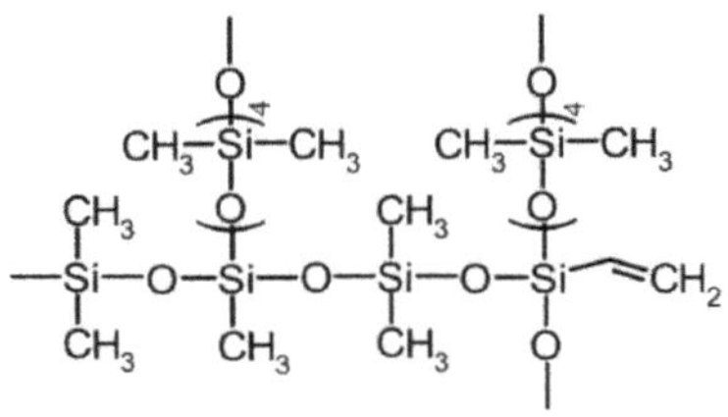

Figure 3. Simplified structure of partially crosslinked silicone resin obtained in synthesis of SIL 1 and SIL 2 dispersions.

Hybrid silicone-acrylic dispersions SIL-ACR 1-A and SIL-ACR 1-B were synthesized by emulsion polymerization of mixtures of acrylic and styrene monomers A and B, respectively, in silicone resin dispersion SIL 1. Hybrid silicone-acrylic dispersions SIL-ACR 2-A and SIL-ACR 2-B were synthesized by emulsion polymerization of mixtures of acrylic and styrene monomers A and B, respectively, in silicone resin dispersion SIL 2. Compositions of acrylic and styrene monomers mixtures A and B corresponded to compositions of acrylic and styrene monomers in starting acrylic/styrene copolymer dispersions ACR A and ACR B. Polymerization was carried out at 78–79 °C for 5 h. After cooling to room temperature, dispersions were neutralized with 25% aqueous NH_3 solution to reach pH ca. 6.0–6.5, then 0.15% of biocide was added and dispersions were filtered through 190 mesh net. Free acrylic and styrene monomers contents as tested by GC were <0.01%. No free VTES or MTES were detected by GC, although small amounts of D4 (ca. 0.4%) and ethanol (ca. 0.1%) were still present. Hybrid acrylic-silicone dispersions ACR-SIL A-1 and ACR-SIL A-2 were synthesized by emulsion polymerization of mixtures of silicone monomers 1 and 2, respectively, in acrylic/styrene copolymer dispersion ACR A. Hybrid acrylic-silicone dispersions (ACR-SIL B-1 and ACR-SIL B-2 were synthesized by emulsion polymerization of mixtures of silicone monomers 1 and 2, respectively, in acrylic/styrene copolymer dispersion ACR B. Compositions of silicone monomers mixtures 1 and 2 corresponded to compositions of silicone monomers in starting silicone resin dispersions SIL 1 and SIL 2. Polymerization was carried out at 88–89 °C for 4 h and then ethanol that was formed in hydrolysis of VTES and MTES was distilled off under vacuum for 3 h. After cooling to room temperature, dispersions were neutralized with 25% NH_3 solution to reach pH ca. 6.0–6.5, then 0.15% of biocide was added and dispersions were filtered through 190 mesh net. No free VTES or MTES or acrylic and styrene monomers were detected by GC in the resulting dispersions, though small amounts of D4 (ca. 0.4%) and ethanol (ca. 0.1%) were still present.

It was essential that the composition and concentration of surfactants remained the same in SIL-ACR and ACR-SIL dispersions, so their properties (and properties of coatings obtained from them) could be compared.

2.3. Characterization of Dispersions

All dispersions were characterized by:

- Solids content, wt.%–percentage of sample mass remaining after drying for 1 h at 80 °C followed by 4 h at 125 °C. The measurements were conducted three times and the mean value was taken.
- pH—using standard indicator paper.
- Viscosity—using Bohlin Instruments CVO 100 rheometer (Cirencester, UK), cone-plate 60 mm diameter and 1° measuring device, shear rate 600 s^{-1}.
- Coagulum content—after filtration of dispersion on 190 mesh net the solids remaining on the net were dried and weighed. Coagulum content (wt.%) was calculated from equation $m_c/m_d \times 100\%$ where m_c was mass of dry coagulum remaining on the net and m_d was mass of dispersion.
- Acrylic and styrene monomers, ethanol and D4 content—by GC (HP 5890 series II apparatus FID detector, Hewlett Packard, Palo Alto, CA, USA)

- Mechanical stability—lack or occurrence of separation during rotation in Hettich Universal 32R centrifuge (Westphalian, The Netherlands) at 4000 r.p.m. for 90 min was considered as good stability.
- Average particle size (nm), particle size distribution and zeta potential (mV)—light-scattering method using Malvern Zeta Sizer apparatus.
- Dispersion particles appearance—transmission electron microscope (TEM) Hitachi 2700 (Tokyo, Japan), dispersions were diluted 1000× with water (1 part of dispersion per 1000 parts of water) for taking pictures. High Angle Annular Dark Field (HAADF) mode also called "Z-contrast" was applied for processing the images reproduced in this paper.
- Minimum film-forming temperature (MFFT)—according to ISO 2115 [34] using Coesfeld apparatus equipped with temperature gradient plate. Temperature range: –3–50 °C.
- Glass transition temperature (T_g) of dispersion solids—by differential scanning calorimetry (DSC) (TA Instruments Q2000 apparatus, New Castle, DE, USA), heat–cool–heat regime, 20 °C/min.

2.4. Characterization of Coatings

Coatings were produced from dispersions by applying them on glass (for testing contact angle, hardness, adhesion or water resistance), aluminium plates (for testing elasticity) or on steel plates (for testing impact resistance and cupping) using 120 μm applicator. Drying was carried out for 30 min at 50 °C and then the coatings were seasoned in a climatic chamber at 23 °C and 55% relative humidity (R.H.) for 72 h. Since no continuous coating could be obtained in this procedure for SIL-ACR 1-B and SIL-ACR 2-B, the relevant dispersions were dried at 8 °C for 2 h and then seasoned as above. The resulting coatings were characterized by:

- Contact angle (water)—according to EN 828:2000, using KRUSS DSA 100E apparatus (KRÜSS GmbH, Hamburg, Germany). The measurements were conducted five times and the mean value was taken.
- Pendulum hardness (Koenig)—according to EN ISO 1522 [35]. The measurements were conducted seven times and the mean value was taken.
- Adhesion—according to EN ISO 2409 [36], the tests were repeated at least three times.
- Elasticity—according to EN ISO 1519 [37], the tests were repeated at least two times.
- Impact resistance (direct and reverse)—according to EN ISO 6272-1 [38], using Erichsen Variable Impact Tester Model 304 (Erichsen, Hemer, Germany). The measurements were conducted at least twice.
- Cupping—according to EN ISO 1520 [39], using Erichsen Cupping Tester (ERICHSEN GmbH & Co. KG, Hemer, Germany). The tests were repeated three times and the mean value was taken.
- Water resistance—glass Petri dishes of 50 mm diameter were filled with distilled water and placed upside-down on the coating, so the coating was covered with 7 mm thick layer of water. Assembles prepared this way were left for 72 h and appearance of coatings was examined for the bubbles size (S0—no bubbles, S2–S5—small to large size of bubbles) and density (0—no bubbles, 2–5 low to high density of bubbles) according to EN ISO 4628-2 [40]. Observation of changes of coating appearance after 6 days under water were also examined.
- Water vapour permeability—according to ASTM F1249 [41]. TotalPerm 063 (Extra Solution) apparatus was used. Tests were conducted at 23 °C for 0.35 mm thick film. Fomblin perfluorinated grease from Solvay Solexis (Brussels, Belgium) was applied to seal the test vessels. The measurements were repeated at least twice.
- Moreover, coatings applied on PET film were examined for surface structure by X-ray photoelectron spectroscopy (XPS)—ULVAC/PHYSICAL ELECTRONICS PHI5000 VersaProbe apparatus (Physical Electronics, Inc., Chigasaki, Japan).

2.5. Characterization of Films

- Percentage swell, i.e., change of the mass caused by soaking in water or organic solvent—ca. 0.12 g samples of film were weighed and placed in 40 mL H_2O or 40 mL toluene contained in closed glass cups and left for 20 h at 23 °C. Then the samples were taken out, delicately dried with filter paper and weighed. Percentage swell was calculated from the equation: % swell = m_1-m_0/m_0 × 100%, where m_0 = mass of the sample before test and m_1 = mass of the sample after test. The tests were repeated three times.

- Mechanical properties (tensile strength and elongation at break)—using Instron 3345 testing machine (Instron, Norwood, MA, USA) according to EN-ISO 527-1 [42] at a pulling rate of 50 mm/min on dumbbell-shaped specimens. The measurements were conducted five times and the mean value was used taken.

3. Results and Discussion

3.1. Properties of Dispersions

Properties of hybrid silicone-acrylic (SIL-ACR) and acrylic-silicone (ACR-SIL) dispersions prepared with SIL/ACR *w/w* 1/3 ratio, starting silicone resin dispersions (SIL 1 and SIL 2) and starting acrylic/styrene copolymer dispersions (ACR A and ACR B), are presented in Table 1. All hybrid dispersions were mechanically stable, slightly opalescent white liquids with low viscosity, pH in the range 5.8–6.3 and solids contents close to 42%. Coagulum content was at a very low level –0.04%–0.38%. Blends of starting SIL and ACR dispersions at *w/w* 1/3 ratio were also made, but the resulting dispersions were not mechanically stable and did not produced continuous coatings at room temperature.

3.1.1. Particle Size and Particle Size Distribution

For hybrid dispersions, particle size distribution was monomodal and rather narrow, although in most cases wider than that for starting SIL and ACR dispersions. Zeta potentials were all very low (i.e., very negative) which indicated good dispersion stability that was confirmed by mechanical stability tests.

The average particle size was distinctly higher for hybrid dispersions SIL-ACR than for starting SIL dispersion and almost the same for ACR-SIL than for starting ACR dispersion (see Figure 4) what could indicate formation of shell on starting SIL dispersion core particles during polymerization of ACR monomers and lack of formation of core-shell particle structure in the case of polymerization of SIL monomers in starting ACR dispersion. The comparison of particle size distribution patterns confirmed that assumption for ACR-SIL dispersions—see Figure 5a. As it can be seen in Figure 5b, in synthesis of SIL-ACR dispersions acrylic/styrene copolymer particles smaller than particles of starting SIL dispersion were probably formed along with core-shell SIL-ACR particles.

In general, average particle size was significantly higher for SIL-ACR dispersions than for ACR-SIL dispersions of the same composition of SIL and ACR parts—see Figure 6 where the particle size distribution of one of the SIL-ACR dispersions (SIL-ACR 2-B) and of the corresponding ACR-SIL dispersion (ACR-SIL B-2) is shown. The reason for that was higher particle size of starting SIL dispersions than for starting ACR dispersions.

Table 1. Properties of hybrid silicone-acrylic (SIL-ACR) and ACR-SIL dispersions and of starting SIL and ACR dispersions. In the case of starting ACR dispersions all properties were determined for dispersions neutralized after polymerization while ACR dispersions before neutralization (with pH ca. 3.0) were used in synthesis of ACR-SIL hybrids. N.A. = Not Applicable because dispersion did not produce continuous film at room temperature (R.T.).

Designation of Dispersions	pH	Solids Content %	Coagulum Content %	Viscosity at 23 °C mPa·s	Average Particle Size nm	Particle Size Distribution nm	Polydispersity	Zeta Potential mV	Minimum Film-Forming Temperature (MFFT) °C	T_g (Disp. Solids) °C
SIL 1	6.3	19.3	0.00	2	121.7	104–157	0.121	−50.9	N.A.	−117.33
SIL 2	6.2	18.5	0.00	3	128.6	116–154	0.086	−49.5	N.A.	−119.46
ACR A	6.2	51.1	<0.1	94	105.7	74–119	0.112	−51.0	+11.4	+17.17
ACR B	6.2	51.5	<0.1	98	112.2	108–135	0.066	−56.0	+32.7	+32.36
SIL-ACR 1-A	5.8	42.2	0.06	20	143.8	69–160	0.074	−57.7	+7.3	−126.66 +17.66
ACR-SIL A-1	6.3	43.2	0.38	140	111.7	71–131	0.096	−55.2	−0.5	−130.54 +14.58
SIL-ACR 1-B	6.3	42.3	0.11	24	140.3	69–190	0.072	−59.5	26.2	−129.09 +30.87
ACR-SIL B-1	6.3	42.0	0.21	61	114.4	98–133	0.071	−55.0	16.0	−126.33 +27.98
SIL-ACR 2-A	6.1	41.7	0.04	20	151.2	107–214	0.064	−59.2	−1.0 [1]	−132.94 +17.29
ACR-SIL A-2	6.3	41.6	0.13	55	109.3	98–122	0.085	−47.2	−0.4	+16.13
SIL-ACR 2-B	6.3	41.4	0.05	16	149.9	125–156	0.057	−55.7	+26.0 [1]	−131.46 +32.56
ACR-SIL B-2	6.3	42.2	0.24	56	115.9	104–130	0.069	−53.6	+14.8	+20.04

[1] For these dispersions also maximum film forming temperature was observed. It was +10.5 °C for SIL-ACR 2-A and +36.2 °C for SIL-ACR 2-B.

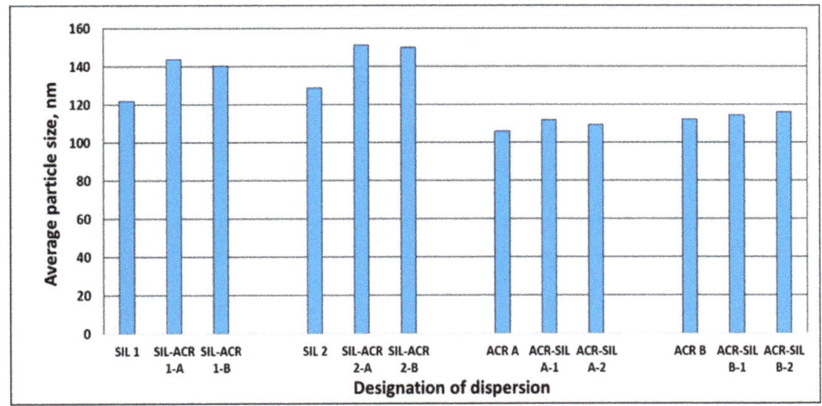

Figure 4. Comparison of average particle size of SIL and ACR dispersions and hybrid SIL-ACR and ACR-SIL dispersions.

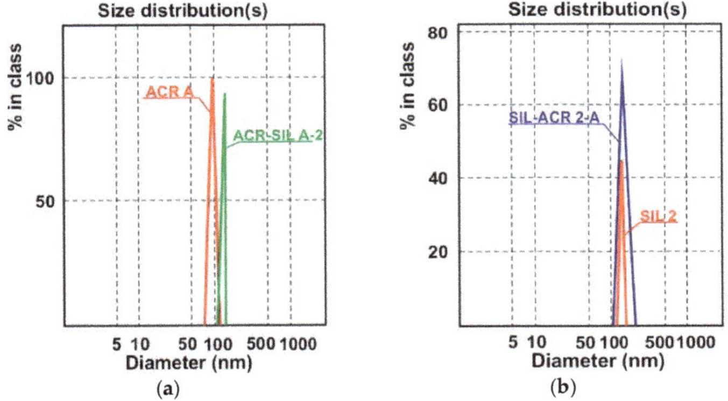

Figure 5. Comparison of particle size distribution patterns of hybrid ACR-SIL A-2 dispersion and starting ACR A dispersion (**a**) and of hybrid SIL-ACR 2-A dispersion and starting SIL 2 dispersion (**b**). X-axis is logarithmic.

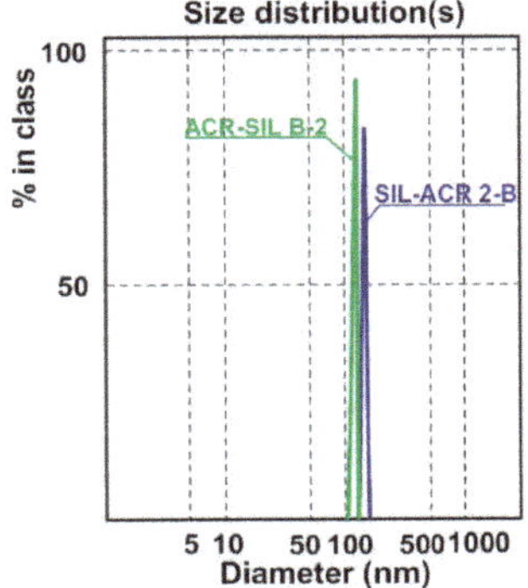

Figure 6. Comparison of particle size distribution patterns for SIL-ACR and ACR-SIL dispersions of the same composition of SIL and ACR parts. X-axis is logarithmic.

3.1.2. Particle Structure

In Figure 7 the structure of hybrid dispersion particles of SIL-ACR and ACR-SIL dispersions determined by TEM is shown. As can be seen in Figure 7a, in the case of SIL-ACR dispersion coalescence of particles proceeded during testing, so the TEM image shows a tiny piece of film rather than the single particle, but it is clear that well defined silicone resin particles (lighter shade) are surrounded by acrylic/styrene copolymer phase (darker shade). Individual particles can be identified better in Figure 7b where lower magnification was used and it can be concluded that a kind of "fruit cake" particle structure where a few "cores" made of one polymer are surrounded by continuous mass of the other polymer was formed during polymerization of ACR monomers in SIL dispersion. In the case of ACR-SIL, dispersion coalescence of particles during testing also proceeded. While both individual particles and aggregates of silicone resin particles and acrylic/styrene copolymer particles were present, it was also possible to identify in TEM images abundant single particles of specific structure shown in Figure 7c. In this structure kinds of spheres made of silicone resin (lighter shade) were embedded in the mass of acrylic/styrene copolymer (darker shade). It can be anticipated that in the course of synthesis of ACR-SIL hybrid dispersions silicone monomers penetrated into acrylic/styrene copolymer particles and after completion of polymerization a kind of sphere of silicone resin was formed because of lack of compatibility of acrylic/styrene copolymer and silicone resin. Such a particle structure called an "embedded sphere" has been found also earlier in polyurethane-acrylic/styrene hybrid dispersions [4].

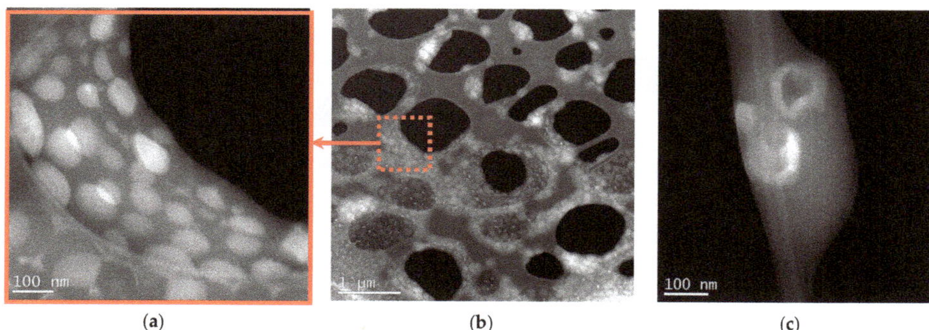

(a)   (b)   (c)

Figure 7. Structure of hybrid dispersion particles settled on the micromesh net as determined by transmission electron microscopy (TEM): (**a**) SIL-ACR dispersion, higher magnification, (**b**) SIL-ACR dispersion, lower magnification, (**c**) ACR-SIL dispersion, higher magnification. Lighter shade represents silicone resin and darker shade–acrylic/styrene copolymer.

Lack of formation of core-shell ACR-SIL hybrid particles in the course of polymerization of silicone monomers in acrylic/styrene copolymer dispersion could have been expected since it was clear from the review of available literature on that subject [11] that only if special approaches were applied to synthesis (e.g., functionalization of acrylic particle surface with silane and hydrolysis of alkoxysilane groups prior to polymerization [22]) the particles with acrylic polymer core and silicone shell could be obtained.

3.1.3. Minimum Film-Forming Temperature (MFFT)

As it can be seen in Figure 8 MFFT values determined for ACR-SIL hybrid dispersions were much lower than for starting ACR dispersion and lower than for SIL-ACR dispersions of the same SIL and ACR parts composition what can be explained by the fact that only a fraction of particles of ACR-SIL dispersion hybrid structure exhibited a hybrid morphology shown in Figure 6b and the presence of separate silicone resin particles resulted in lower MFFT.

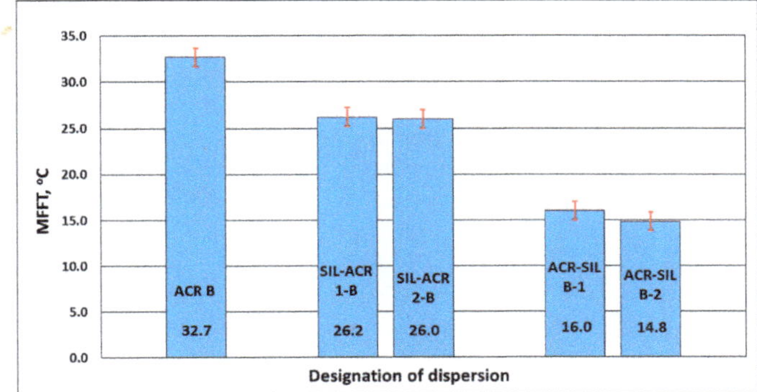

Figure 8. Comparison of MFFT determined for hybrid dispersions SIL-ACR and ACR-SIL. MFFT of starting ACR dispersion used for synthesis of ACR-SIL dispersions is also shown.

3.1.4. Glass Transition Temperature (T_g)

DSC results showed that hybrid dispersion solids usually exhibited two T_{gs}: one corresponding to SIL part at c.a. -120 °C and the other corresponding to ACR part in the range of ca. 15–30 °C, depending on the T_g of starting acrylic/styrene copolymer—see Figure 9 where DSC patterns determined for starting SIL and ACR dispersions and for SIL-ACR and ACR-SIL dispersions having the same composition of ACR and SIL parts are presented.

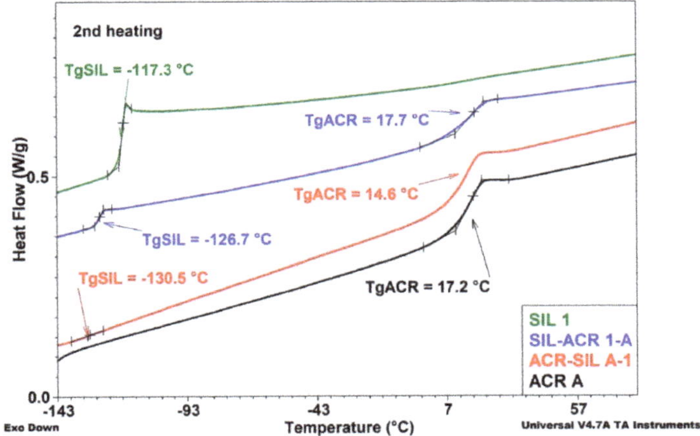

Figure 9. Differential scanning calorimetry (DSC) patterns determined for starting SIL and ACR dispersions and for SIL-ACR and ACR-SIL dispersions having the same composition of ACR and SIL parts.

Only for two dispersions (ACR-SIL A-2 and ACR-SIL B-2) just one T_g was detected at around 16 and 20 °C, respectively, what suggested that in the case of these two dispersions the particle structure was rather uniform and no separate silicone resin particles were formed. This phenomenon can be explained by the fact that in these two dispersions silicone monomers (D4 + ethoxy-functional silane) mixture that was polymerized in acrylic/styrene copolymer dispersion contained much more VTES (more polar) and did not contain MTES (less polar), so penetration into acrylic/styrene copolymer particles was easier and grafting of VTES on acrylic/styrene copolymer and formation of "embedded sphere" structures shown in Figure 7c were much more probable.

It was also interesting that T_{gs} of SIL and ACR parts of all hybrid dispersion solids where two glass transitions were detected were significantly lower than T_{gs} of starting SIL and ACR dispersions solids. Decrease in T_g of ACR part can be explained by plasticizing effect of modification with silicone resin. However, in order to clarify why decrease in T_g of SIL partly occurred, more insight is needed to the processes which took place in the course of both silicone monomers polymerization in acrylic/styrene copolymer dispersion and acrylic/styrene monomers polymerization in silicone resin dispersion. The key assumption (confirmed by the hybrid particle structures) is that in hybrid dispersion particles silicone resin particles are "trapped" within a mass of acrylic/styrene copolymer, so D4 and higher oligodimethylsiloxane cycles (e.g., D5) which are always present in SIL dispersions [43] and are also formed in synthesis of hybrid ACR-SIL dispersions are also "trapped" and therefore cannot be released during drying and may plasticize the silicone resin contained in dispersion solids. That "trapping" of silicone resin in acrylic/styrene copolymer part of hybrid dispersion particles should be more distinct if the SIL part contained more VTES because of possibility of grafting the decrease in T_g should be more distinct for hybrid dispersions ACR-SIL 1-A and ACR-SIL 1-B than for hybrid dispersions ACR-SIL 2-A and ACR-SIL 2-B. Comparison of the relevant T_g values in Table 1 confirmed that this was actually the case.

3.2. Properties of Coatings and Films

Properties of coatings and films obtained from hybrid silicone-acrylic (SIL-ACR) and acrylic-silicone (ACR-SIL) dispersions prepared with SIL/ACR w/w 1/3 ratio, starting silicone resin dispersions (SIL 1 and SIL 2) and starting acrylic/styrene copolymer dispersions (ACR A and ACR B) are presented in Table 2. Some hybrid dispersions and starting silicone resin dispersions did not form mechanically strong continuous coatings or films, but certain properties like e.g., contact angle or % swell could be determined by casting layers which, after drying, formed mechanically weak coatings or films.

It is essential that for all hybrid dispersions the key coating properties that were expected to improve as compared to acrylic/styrene copolymer dispersions (contact angle, water vapor permeability and water resistance) actually did improve significantly. Mechanical properties of coatings (e.g., impact resistance or elasticity) also improved, but hardness decreased what could be expected. The same trend was reflected in film properties—increase in elongation at break was accompanied by a decrease in tensile strength.

3.2.1. Surface Properties

The high contact angle of coatings is important since it means high surface hydrophobicity and, consequently, lower water uptake and lower dirt deposition [5]. As can be seen in Table 2, all coatings obtained from hybrid SIL-ACR and ACR-SIL dispersions showed high contact angles in the range of 80–90° while contact angles recorded for films obtained from starting ACR dispersions were quite low (ca. 30°). It is worth to note that contact angles recorded for coatings produced from ACR-SIL dispersions were generally higher than those recorded for coatings produced from SIL-ACR dispersions (see Figure 10) what indicates that in the former case more silicone migrated to the coating surface.

Migration of silicone to the coating surface observed for coatings containing silicones was described in the earlier papers, e.g., [32,44,45] and was fully confirmed by XPS also for coatings obtained from SIL-ACR and ACR-SIL hybrid dispersions. In Figure 11 the percentage of Si in the layers close to coating surface as determined by XPS for hybrid SIL-ACR and ACR-SIL dispersions is plotted against distance from the surface. It is clear from Figure 11 that in the coatings obtained from hybrid dispersions silicone migrated to coating surface and that migration was different for coatings obtained from ACR-SIL dispersions than for those obtained from SIL-ACR dispersions, most probably due to "trapping" of silicone resin in acrylic/styrene copolymer particles in the latter coating.

Table 2. Properties of coatings and films made from hybrid SIL-ACR and ACR/SIL dispersions and of starting SIL and ACR dispersions. N.A. = Not Applicable because dispersion did not produce continuous coating or/and film at R.T. Although SIL-1 and SIL-2 dispersions did not produce continuous a mechanically strong coatings, it was possible to measure their contact angles on very mechanically weak coats that were cast on glass Petri dishes.

Designation of Dispersions	Contact Angle (H_2O) (°)	Water Vapour Permeability $g/m^2/24h$	Water Resistance after 72 h	Impact Resistance (direct) J	Impact Resistance (reverse) J	Cupping mm	Elasticity (Rod Diameter 2 mm)	Hardness (Koenig)	Adhesion to Glass	Swell in H_2O %	Swell in Toluene %	Tensile Strength MPa	Elongation at Break %
SIL 1	111				N.A.					18	202		N.A.
SIL 2	104				N.A.					10	387		N.A.
ACR A	30	28.1	>5(S5) 5(S2)	2.0	19.6	10.7	passed	0.082	5	14	1156	4.2	1000
ACR B	35	15.6	Medium whitening 5(S2)	2.0	0	10.7	failed	0.458	5	15	1547	12.0	340
SIL-ACR 1-A	83	56.5	Medium whitening 0(S0)	9.8	19.6	11.6	passed	0.040	3	26	591	2.1	773
ACR-SIL A-1	95	45	Light whitening 5(S2)	15.7	19.6	11.0	passed	0.022	2	21	1050	0.8	1851
SIL-ACR 1-B	81	64.5	Light whitening 0(S0)	0	0	10.9	passed	0.085	5	11	561	4.3	11
ACR-SIL B-1	92	34.6	Medium whitening	5.9	19.6	10.9	passed	0.050	5	26	905	3.1	1015
SIL-ACR 2-A	77				N.A.					20	605		
ACR-SIL A-2	92	39.4	0(S0) Medium whitening	13.7	19.6	10.5	passed	0.034	5	16	1112	0.9	1516
SIL-ACR 2-B	85				N.A.					9	665		N.A.
ACR-SIL B-2	82	28.0	0(S0) Medium whitening	3.9	19.6	11.0	passed	0.058	5	31	991	3.2	947

75

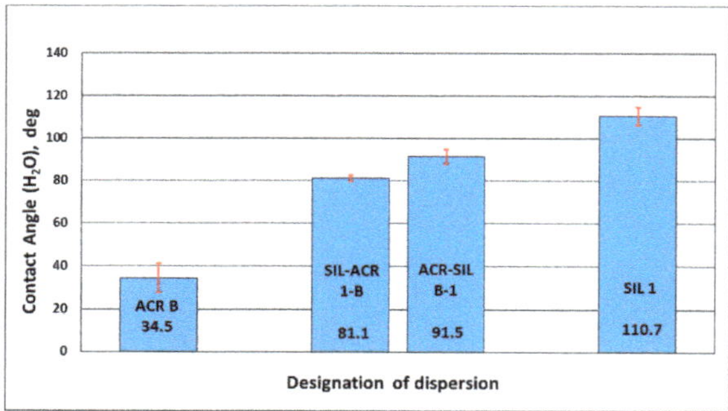

Figure 10. Comparison of contact angle values determined for coatings obtained from starting SIL and ACR dispersions and corresponding hybrid SIL-ACR and ACR-SIL dispersions having the same composition of ACR and SIL parts. Contact angle determined for starting acrylic/styrene copolymer dispersion (ACR B) is also shown.

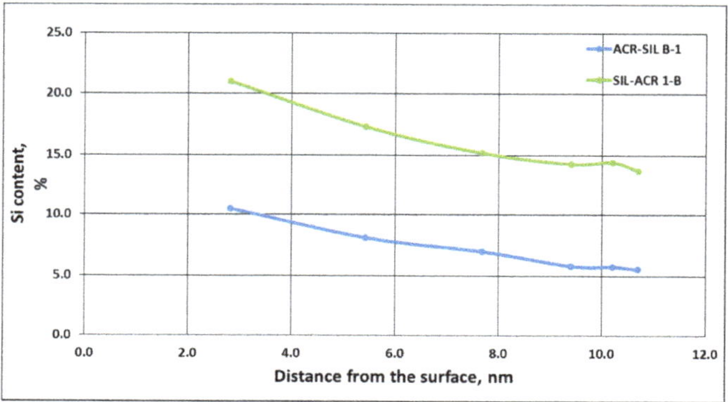

Figure 11. Decrease in Si content with distance from coating surface determined by XPS for coatings obtained from SIL-ACR and ACR-SIL dispersions.

3.2.2. Water Resistance

Good water resistance of architectural paints is crucial since it ensures longer life of the paint and better comfort of the building walls (lack of water uptake) if combined with high water vapour permeability. Therefore, determination of the water resistance of coatings produced from dispersions which are intended to be applied as binders for architectural paints seems to be very important test. In our investigations we measured water resistance of coatings obtained from starting ACR dispersions and from SIL-ACR and ACR-SIL dispersions using our own method partly described in Section 2.4 and the results were assessed based on EN ISO 4628-2 [40]. All coatings made from hybrid dispersions exhibited better water resistance than those produced from starting ACR dispersions and it was significantly better for coatings obtained from ACR-SIL dispersions than from SIL-ACR dispersions—see Figure 12 where photos of coatings produced from different dispersions and left under water for 6 days are shown.

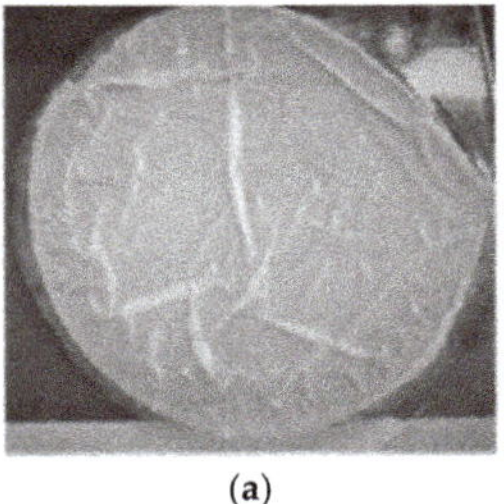
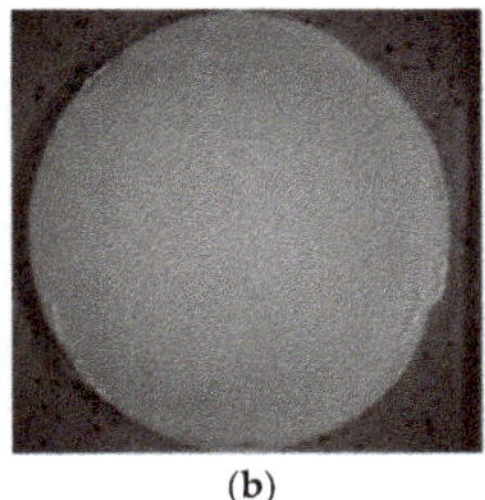
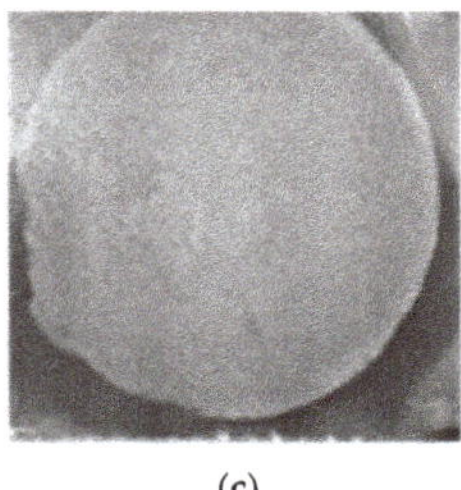

(a)   (b)   (c)

Figure 12. Comparison of water resistance of starting acrylic/styrene copolymer dispersion. ACR A (**a**), acrylic-silicone dispersion (ACR-SIL A-1) (**b**) and corresponding silicone-acrylic dispersion SIL-ACR (**c**). Samples were kept under water for 6 days. ACR A—deterioration of coating occurred, ACR-SIL A-1—coating did not change except for light whitening, SIL-ACR 1-A—coating changed significantly– numerous small bubbles.

3.2.3. Swell in Water and in Toluene

As can be concluded from Table 2 percent of swell in water was very similar for all films (despite of differences in water resistance of coatings) and was quite low (ca. 20%) while swell in toluene that can be considered as a measure of crosslinking density (higher swell means lower crosslinking density) was much higher for films made from ACR dispersions than for films made from SIL dispersions, and also much higher for films made from hybrid ACR-SIL dispersions than for films made from SIL-ACR dispersions—see the relevant comparison in Figure 13.

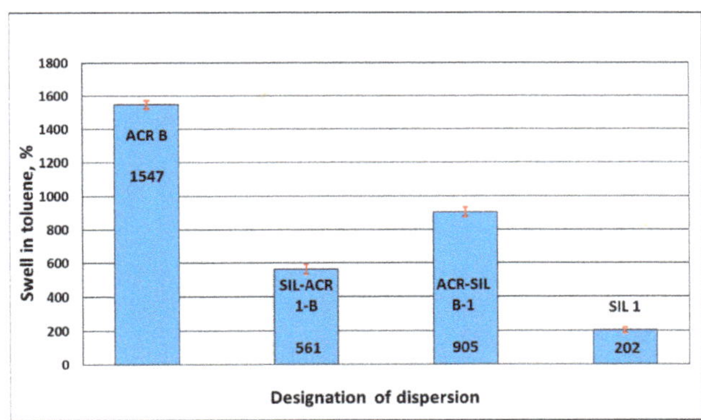

Figure 13. Comparison of % swell in toluene determined for starting silicone resin dispersion (SIL-1), starting acrylic/styrene copolymer dispersion (ACR B) and hybrid dispersions ACR-SIL B-1 and SIL-ACR 1-B having the same composition of SIL and ACR parts.

The difference between crosslinking density of films (i.e., also for coatings) made from ACR-SIL and SIL-ACR dispersions having the same composition of ACR and SIL parts can be explained by a higher possibility of grafting of acrylic/styrene monomers on silicone resin than of grafting VTES on acrylic/styrene copolymer. Another reason can be a higher possibility of trapping of partly crosslinked silicone resin inside particles made of acrylic/styrene copolymer in the case of films made from SIL-ACR dispersions than in the case of films made from ACR-SIL dispersions—see the discussion of hybrid dispersions particle structures contained in Section 3.1.2.

3.2.4. Water Vapour Permeability

As has already been pointed out in Section 3.2.2, good architectural paint should exhibit not only good water resistance, but also good water vapour permeability. This positive combination of properties can be achieved in practice only for paints based on silicone-acrylic binders because silicone polymers are characterized by good permeability of gases due to high mobility of poly(dimethylsiloxane) chains. It was proved in our study that coatings produced from hybrid SIL-ACR and ACR-SIL dispersions showed higher water vapour permeability than those produced from starting ACR dispersions—see Figure 14.

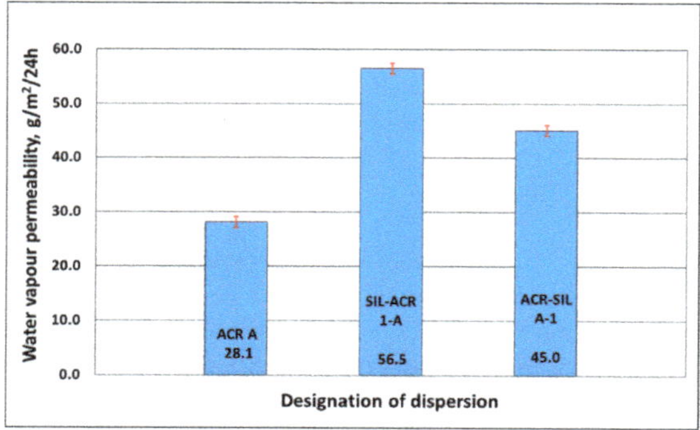

Figure 14. Comparison of water vapour permeability determined for starting acrylic/styrene copolymer dispersion (ACR A) and hybrid dispersions ACR-SIL A-1 and SIL-ACR 1-A having the same composition of SIL and ACR parts.

It can be noted from the results presented in Figure 14 that water vapour permeability was better for coatings obtained from SIL-ACR dispersions than from ACR-SIL dispersions, probably because of differences in coating structure that resulted from differences in dispersion particle structure.

3.2.5. Mechanical Properties

If the results of testing the mechanical properties of coatings and films produced from hybrid ACR-SIL and SIL-ACR dispersions presented in Table 2 are compared with mechanical properties of coatings produced from starting ACR dispersions, it is clear that modification with silicone led generally to less brittle coatings, especially in the case of starting dispersion ACR A. The most spectacular difference was in the (direct) impact resistance of coatings—see Figure 15.

For coatings and films produced from starting dispersion ACR B and hybrid coatings and films where ACR B composition of monomers was applied in synthesis of the relevant dispersions, the results of mechanical tests were much less convincing, presumably because T_g of ACR B was quite high (over 30 °C). Cupping test results were good for all coatings and in direct elasticity measurements, only coatings produced from starting dispersion ACR B failed. Elongation at break increased for some films made from hybrid dispersions as compared to films made from starting ACR dispersions and decreased for some others (specifically for these produced from hybrid dispersions with particles having ACR B composition of ACR part) and tensile strength decreased for all films where this could be expected taking into account plasticizing effect of silicone resin. Much higher elongation at break and much lower tensile strength observed for films made from ACR-SIL dispersions than from SIL-ACR dispersions can be explained by a different supramolecular structure of films that results from different morphology of hybrid dispersion particles (see Figure 7) that coalesce to produce these films in the process of air-drying of dispersions.

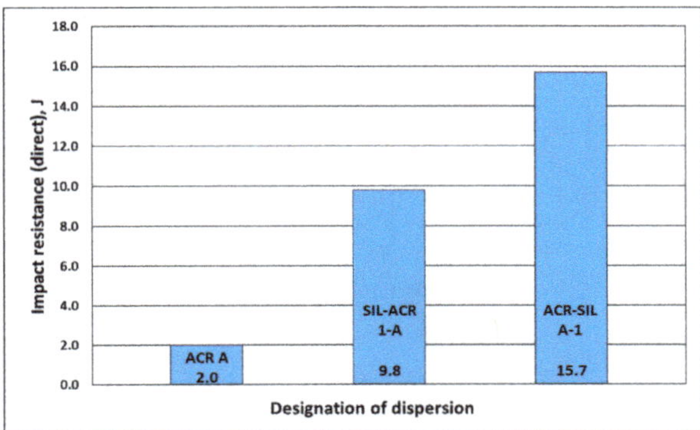

Figure 15. Comparison of impact resistance (direct) determined for coatings obtained from starting acrylic/styrene copolymer dispersion (ACR A) and hybrid dispersions ACR-SIL A-1 and SIL-ACR 1-A having the same composition of SIL and ACR parts.

4. Conclusions

Simultaneous synthesis of aqueous silicone-acrylic and acrylic-silicone hybrid dispersions (SIL-ACR and ACR-SIL) by (1) emulsion polymerization of acrylic/styrene monomers (BA, ST, KA and MA) mixtures of different composition (ACR A and ACR B) in aqueous dispersions of silicone resins of different composition (SIL 1 and SIL 2) and (2) emulsion polymerization of silicone monomers (D4, VTES and MTES) mixtures of different composition (SIL 1 and SIL 2) in aqueous dispersions of acrylic/styrene copolymers (ACR A and ACR B) was successfully conducted. Hybrid dispersions had good mechanical stability, low minimum film-forming temperature and particle size in the range of 100–150 nm, narrow particle size distribution, and contained very little of coagulate. TEM investigation of hybrid dispersions particle structure revealed that particles of SIL-ACR dispersions exhibited "fruit cake" structure while particles of ACR-SIL dispersions showed "embedded sphere" structure. For most of the dispersions two separate T_{gs} of dispersion solids (one for SIL part and the other for ACR part) that were detected by DSC were lower than T_{gs} of corresponding starting SIL and ACR dispersions while single T_g was detected for two of them. These differences were explained by differences in dispersion particle structure.

Most of the hybrid dispersions formed mechanically strong continuous coatings and films. As compared to coatings obtained from starting ACR dispersions, those obtained from hybrid dispersions showed much higher contact angles, much better water resistance and water vapour permeability and exhibited much better impact resistance. Different coating properties were observed when coatings were produced from SIL-ACR and ACR-SIL dispersions having the same composition of ACR and SIL parts, which most probably resulted from different structure of dispersions particles. Films produced from hybrid dispersions were less brittle than those produced from starting ACR dispersions. Determinations of % swell in toluene measured for films produced from hybrid dispersions revealed the difference between crosslinking density of films (i.e., also for coatings) made from ACR-SIL and SIL-ACR dispersions having the same composition of ACR and SIL parts, which was explained by higher possibility of grafting of acrylic/styrene monomers on silicone resin than of grafting VTES on acrylic/styrene copolymer. The authors believe that the selected hybrid dispersions described in this paper can be applied as binders in the formulation of architectural paints that will be characterized by high water resistance and high surface hydrophobicity combined with high water vapour permeability.

Author Contributions: Conceptualization, J.K. and J.T.; Methodology, J.T., W.D., I.O.-K., A.K., K.S. and J.P.; Investigation, J.T., W.D. and A.K.; Writing-Original Draft Preparation, J.K.; Writing-Review & Editing, J.T. and W.D.; Visualization, J.P., W.D. and J.T.; Supervision, J.K., D.W. and M.W.; Project Administration and Funding Aquisition, J.K.

Funding: This research was funded by Polish State R&D Centre (NCBiR, No. PBS/B1/8/2015).

Acknowledgments: The authors wish to thank Piotr Bazarnik from Warsaw University for conducting TEM studies, and Janusz Sobczak from Polish Academy of Sciences for conducting XPS studies. The assistance of Colleagues from the Industrial Chemistry Research Institute in testing mechanical properties of films, water vapour permeability and dispersion particle size distribution, is also acknowledged.

Conflicts of Interest: The authors declare no conflict of interest.

References

1. Rodríguez, R.; de Las Heras Alarcón, C.; Ekanayake, P.; McDonald, P.J.; Keddie, J.L.; Barandiaran, M.J.; Asua, J.M. Correlation of silicone incorporation into hybrid acrylic coatings with the resulting hydrophobic and thermal properties. *Macromolecules* **2008**, *41*, 8537–8546. [CrossRef]
2. Kickelbick, G. Introduction to hybrid materials. In *Hybrid Materials: Synthesis, Characterization and Applications*, 1st ed.; Wiley-VCH Verlag GmbH & Co. KGaA: Weinheim, Germany, 2007; pp. 1–46.
3. Castelvetro, V.; de Vita, C. Nanostructured hybrid materials from aqueous polymer dispersions. *Adv. Colloid Interface Sci.* **2004**, *108–109*, 167–185. [CrossRef]
4. Kozakiewicz, J.; Koncka-Foland, A.; Skarżyński, J.; Legocka, I. Synthesis and characterization of aqueous hybrid polyurethane-urea-acrylic/styrene polymer dispersions. In *Advances in Urethane Science and Technology*, 1st ed.; Klempner, D., Frisch, K.C., Eds.; RAPRA Technology Ltd.: Shrewsbury, UK, 2001; pp. 261–334.
5. Khanjani, J.; Pazokifard, S.; Zohuriaan-Mehr, M.J. Improving dirt pickup resistance in waterborne coatings using latex blends of acrylic/PDMS polymers. *Prog. Org. Coat.* **2017**, *102*, 151–166. [CrossRef]
6. Peruzzo, P.J.; Anbinder, P.S.; Pardini, O.R.; Vega, J.; Costa, C.A.; Galembeck, F.; Amalvy, J.I. Waterborne polyurethane/acrylate: Comparison of hybrid and blend systems. *Prog. Org. Coat.* **2011**, *72*, 429–437. [CrossRef]
7. Ma, J.Z.; Liu, Y.H.; Bao, Y.; Liu, J.L.; Zhang, J. Research advances in polymer emulsion based on "core-shell" structure design. *Adv. Colloid Interface Sci.* **2013**, *197–198*, 118–131. [CrossRef]
8. Mittal, V. *Advanced Polymer Nanoparticles: Synthesis and Surface Modification*, 1st ed.; CRC Press: Boca Raton, FL, USA, 2010.
9. Guyot, A.; Landfester, K.; Schork, F.J.; Wang, C. Hybrid polymer latexes. *Prog. Polym. Sci.* **2007**, *32*, 1439–1461. [CrossRef]
10. Ghosh Chaudhuri, R.; Paria, S. Core/shell nanoparticles: Classes, properties, synthesis mechanisms, characterization, and applications. *Chem. Rev.* **2012**, *112*, 2373–2433. [CrossRef]
11. Kozakiewicz, J.; Ofat, I.; Trzaskowska, J. Silicone-containing aqueous polymer dispersions with hybrid particle structure. *Adv. Colloid Interface Sci.* **2015**, *223*, 1–39. [CrossRef]
12. Holmes, D. Controlling the morphology of composite latex particles. *Inquiry J.* **2005**, *4*. Available online: https://scholars.unh.edu/inquiry_2005/4/ (accessed on 25 December 2018).
13. Winzor, C.L.; Sundberg, D.C. Conversion dependent morphology predictions for composite emulsion polymers: 1. Synthetic lattices. *Polymer* **1992**, *33*, 3797–3809. [CrossRef]
14. Chen, Y.C.; Dimonie, V.; El-Aasser, M.S. Interfacial phenomena controlling particle morphology of composite latexes. *J. Appl. Polym. Sci.* **1991**, *42*, 1049–1063. [CrossRef]
15. Ferguson, C.J.; Russel, G.T.; Gilbert, R.G. Modelling secondary particle formation in emulsion polymerisation: Application to making core–shell morphologies. *Polymer* **2002**, *43*, 4557–4570. [CrossRef]
16. Sundberg, D.S.; Durant, Y.G. Latex particle morphology, fundamental aspects: A review. *Polym. React. Eng.* **2003**, *11*, 379–432. [CrossRef]
17. Eduok, U.; Faye, O.; Szpunar, J. Recent developments and applications of protective silicone coatings: A review of PDMS functional materials. *Prog. Org. Coat.* **2017**, *111*, 124–163. [CrossRef]
18. He, W.D.; Cao, C.T.; Pan, C.Y. Formation mechanism of silicone rubber particles with core-shell structure by seeded emulsion polymerization. *J. Appl. Polym. Sci.* **1996**, *61*, 383–388. [CrossRef]

19. He, W.D.; Pan, C.Y. Influence of reaction between second monomer and vinyl group of seed polysiloxane on seeded emulsion polymerization. *J. Appl. Polym. Sci.* **2001**, *80*, 2752–2758. [CrossRef]

20. Lin, M.; Chu, F.; Gujot, A.; Putaux, J.L.; Bourgeat-Lami, E. Silicone-polyacrylate composite latex particles. Particles formation and film properties. *Polymer* **2005**, *46*, 1331–1337. [CrossRef]

21. Shen, J.; Hu, Y.; Li, L.X.; Sun, J.W.; Kan, C.Y. Fabrication and characterization of polysiloxane/polyacrylate composite latexes with balanced water vapor permeability and mechanical properties: Effect of silane coupling agent. *J. Coat. Technol. Res.* **2018**, *15*, 165–173. [CrossRef]

22. Kan, C.Y.; Kong, X.Z.; Yuan, Q.; Liu, D.S. Morphological prediction and its application to the synthesis of polyacrylate/polysiloxane core/shell latex particles. *J. Appl. Polym. Sci.* **2001**, *80*, 2251–2258. [CrossRef]

23. Bourgeat-Lami, E.; Tissot, I.; Lefevbre, F. Synthesis and characterization of SiOH-functionalized polymer latexes using methacryloxy propyl trimethoxysilane in emulsion polymerization. *Macromolecules* **2002**, *35*, 6185–6191. [CrossRef]

24. Kozakiewicz, J.; Rościszewski, P.; Rokicki, G.; Kołdoński, G.; Skarżyński, J.; Koncka-Foland, A. Aqueous dispersions of siloxane-acrylic/styrene copolymers for use in coatings–preliminary investigations. *Surf. Coat. Int. Part B* **2001**, *84*, 301–307. [CrossRef]

25. Kan, C.Y.; Zhu, X.L.; Yuan, Q.; Kong, X.Z. Graft emulsion copolymerization of acrylates and siloxane. *Polym. Adv. Technol.* **1997**, *8*, 631–633. [CrossRef]

26. Li, W.; Shen, W.; Yao, W.; Tang, J.; Xu, J.; Jin, L.; Zhang, J.; Xu, Z. A novel acrylate-PDMS composite latex with controlled phase compatibility prepared by emulsion polymerization. *J. Coat. Technol. Res.* **2017**, *14*, 1259–1269. [CrossRef]

27. Xu, W.; An, Q.; Hao, L.; Zhang, D.; Zhang, M. Synthesis and characterization of self-crosslinking fluorinated polyacrylate soap-free lattices with core-shell structure. *Appl. Surf. Sci.* **2013**, *268*, 373–380. [CrossRef]

28. Hao, G.; Zhu, L.; Yang, W.; Chen, Y. Investigation on the film surface and bulk properties of fluorine and silicon contained polyacrylate. *Prog. Org. Coat.* **2015**, *85*, 8–14. [CrossRef]

29. Li, J.; Zhong, S.; Chen, Z.; Yan, X.; Li, W.; Yi, L. Fabrication and properties of polysilsesquioxane-based trilayer core-shell structure latex coatings with fluorinated polyacrylate and silica nanocomposite as the shell layer. *J. Coat. Technol. Res.* **2018**, *15*, 1077–1078. [CrossRef]

30. Xu, W.; Hao, L.; An, Q.; Wang, X. Synthesis of fluorinated polyacrylate/polysilsesquioxane composite soap-free emulsion with partial trilayer core-shell structure and its hydrophobicity. *J. Polym. Res.* **2015**, *22*, 20. [CrossRef]

31. Kozakiewicz, J.; Ofat, I.; Legocka, I.; Trzaskowska, J. Silicone-acrylic hybrid aqueous dispersions of core–shell particle structure and corresponding silicone-acrylic nanopowders designed for modification of powder coatings and plastics. Part I–Effect of silicone resin composition on properties of dispersions and corresponding nanopowders. *Prog. Org. Coat.* **2014**, *77*, 568–578. [CrossRef]

32. Kozakiewicz, J.; Ofat, I.; Trzaskowska, J.; Kuczynska, H. Silicone-acrylic hybrid aqueous dispersions of core–shell particle structure and corresponding silicone-acrylic nanopowders designed for modification of powder coatings and plastics. Part II: Effect of modification with silicone-acrylic nanopowders and of composition of silicone resin contained in those nanopowders on properties of epoxy-polyester and polyester powder coatings. *Prog. Org. Coat.* **2015**, *78*, 419–428. [CrossRef]

33. Pilch-Pitera, B.; Kozakiewicz, J.; Ofat, I.; Trzaskowska, J.; Spirkova, M. Silicone-acrylic hybrid aqueous dispersions of core–shell particle structure and corresponding silicone-acrylic nanopowders designed for modification of powder coatings and plastics. Part III: Effect of modification with selected silicone-acrylic nanopowders on properties of polyurethane powder coatings. *Prog. Org. Coat.* **2015**, *78*, 429–436. [CrossRef]

34. *ISO 2115 Polymer Dispersions–Determination of White Point Temperature and Minimum Film-Forming Temperature*; International Organization for Standardization: Geneva, Switzerland, 1996.

35. *ISO 1552 Paints and Varnishes–Pendulum Damping Test*; International Organization for Standardization: Geneva, Switzerland, 2006.

36. *ISO 2409 Paints and Varnishes–Cross-Cut Test*; International Organization for Standardization: Geneva, Switzerland, 2013.

37. *ISO 1519 Paints and Varnishes–Bend Test (Cylindrical Mandrel)*; International Organization for Standardization: Geneva, Switzerland, 2011.

38. *ISO 6272-1 Paints and Varnishes–Rapid-Deformation (Impact Resistance) Tests–Part 1: Falling-Weight Test, Large-Area Indenter*; International Organization for Standardization: Geneva, Switzerland, 2011.

39. *ISO 1520 Paints and Varnishes–Cupping Test*; International Organization for Standardization: Geneva, Switzerland, 2006.

40. *ISO 4628-2 Paints and Varnishes–Evaluation of Degradation of Coatings–Designation of Quantity and Size of Defects, and of Intensity of Uniform Changes in Appearance–Part 2: Assessment of Degree of Blistering*; International Organization for Standardization: Geneva, Switzerland, 2016.

41. *ASTM F1249 Standard Test Method for Water Vapor Transmission Rate Through Plastic Film and Sheeting Using a Modulated Infrared Sensor*; ASTM International: West Conshohocken, PA, USA, 2013.

42. *ISO 527-1 Plastics–Determination of Tensile Properties–Part 1: General Principles*; International Organization for Standardization: Geneva, Switzerland, 2012.

43. Liu, Y. *Silicone Dispersions*, 1st ed.; CRC Press: Boca Raton, FL, USA, 2016.

44. Mequanint, K.; Sanderson, R. Self-assembling of metal coatings from phosphate and siloxane-modified polyurethane dispersions: An analysis of the coating interface. *J. Appl. Polym. Sci.* **2003**, *88*, 893–899. [CrossRef]

45. Ofat, I.; Kozakiewicz, J. Modification of epoxy-polyester and polyester powder coatings with silicone-acrylic nanopowders–effect on surface properties of coatings. *Polimery* **2014**, *59*, 643–649. [CrossRef]

 *coatings*

Article

Long-Term Hydrolytic Degradation of the Sizing-Rich Composite Interphase

Andrey E. Krauklis *, Abedin I. Gagani and Andreas T. Echtermeyer

Department of Mechanical and Industrial Engineering, Norwegian University of Science and Technology,
7491 Trondheim, Norway; abedin.gagani@ntnu.no (A.I.G.); andreas.echtermeyer@ntnu.no (A.T.E.)
* Correspondence: andrejs.krauklis@ntnu.no or andykrauklis@gmail.com; Tel.: +371-268-10-288

Received: 1 April 2019; Accepted: 17 April 2019; Published: 19 April 2019

Abstract: Glass fiber-reinforced composites are exposed to hydrolytic degradation in subsea and offshore applications. Fiber-matrix interphase degradation was observed after the matrix was fully saturated with water and typical water absorption tests according to ASTM D5229 were stopped. Due to water-induced dissolution, fiber-matrix interphase flaws were formed, which then lead to increased water uptake. Cutting sample plates from a larger laminate, where the fibers were running parallel to the 1.5 mm long short edge, allowed the hydrolytic degradation process to be studied. The analysis is based on a full mechanistic mass balance approach considering all the composite's constituents: water uptake and leaching of the matrix, dissolution of the glass fibers, and dissolution of the composite interphase. These processes were modeled using a combination of Fickian diffusion and zero-order kinetics. For the composite laminate studied here with a saturated epoxy matrix, the fiber matrix interphase is predicted to be fully degraded after 22 to 30 years.

Keywords: composites; sizing; interphase; glass fibers; environmental degradation; aging; model; kinetics; durability; hydrolysis

---

1. Introduction

Fiber-reinforced polymer (FRP) composites have experienced a rapid rise in use in the past 50 years due to their high strength, stiffness, relatively light weight and good corrosion resistance, especially when compared with more traditional structural materials such as steel and aluminum [1]. The reason for such superior performance is the synergistic interaction between the constituent materials inside the composite [1]. One such material is the sizing, which is a multi-component coating on the surface of the fibers. During the manufacture of FRPs, this results in the formation of a sizing-rich composite interphase between the reinforcing fibers and the matrix polymer [2]. This composite interphase is of vital importance since the mechanical properties of composite materials are often determined by whether the mechanical stresses can be efficiently transferred from the matrix to the reinforcing fibers [3–5]. The quality of the interfacial interaction is strongly dependent on the adhesional contact and the presence of flaws in the interphase [6]. It is generally agreed that the composite interphase is often the mechanical weak link and a potential source for the initiation of defects in fiber-reinforced composite structures [5].

Composite laminates are often exposed to aqueous and humid environments. Environmental aging is especially interesting for marine, offshore and deep-water applications of composites, such as oil risers and tethers [7–12]. It has been reported that water and humid environments negatively impact the mechanical properties of FRPs partially because of a loss of the interfacial bonding [5,12–15]. Flaws in the interphase can be introduced due to the interaction of the interphase with water taken up from the environment [6]. The removal of the sizing material can also lead to a microcrack initiation at the surface of glass fibers. Furthermore, various sizing components can be extracted by water, resulting in the loss of the material [16–20]. Quantifying the water-induced aging is especially important for glass fiber-reinforced composites since the glass fibers are highly hygroscopic [5]. The environmental durability is one of the limiting factors in the structural applications [21], since the superior strength and stiffness of such materials are often compromised by the uncertainty of the material's interaction with the environment [22]. Durability is a primary issue because environmental factors such as moisture, temperature and the state of stress to which the material is exposed can degrade interfacial adhesion as well as the properties of the constituent phases. Environmental aging is mainly important at high temperatures, since the dissolution reactions are accelerated at higher temperatures. Therefore, it is of great importance to understand the environmental aging and dissolution kinetics of a sizing-rich composite interphase.

1.1. Sizing and its Composition

The sizing which forms the interphase, has typically a proprietary composition. Available information about commercial glass fibers tends to contain only one or two sizing-related details. The first is an indication of the chemical compatibility of the sizing with the matrix polymer, e.g., epoxy, as in this case. The second is a value for the loss on ignition (LOI), which indicates the amount of sizing [23]. The key functions of the sizing are: (1) to protect the glass fibers during handling and production; (2) to ensure a high level of stress transfer capability across the fiber-matrix interphase; and (3) to protect the composite matrix interphase against environmental degradation [12].

A typical sizing consists of about 20 chemicals. The most important chemical is an organofunctional silane commonly referred to as a coupling agent [24–26], which is the main component that promotes adhesion and stress-transfer between the polymer matrix and the fiber [12]. It also provides improvements in the interphase strength and hygrothermal resistance of the composite interphase [26–28]. The silane coupling agents have the general structure [X–Si(–O–R)$_3$] where R is a methyl or ethyl group and X is a reactive group towards the polymer, in this case, an amine group. When applied to fibers, a silane coupling agent is first hydrolyzed to a silanol in presence of water. It is unstable and further condenses onto the fibers by producing a siloxane/poly(siloxane) network, which then partially becomes covalently bonded to the glass fiber surface. During the composite manufacture, the X reactive groups of the silane may react with the thermosetting matrix polymer, leading to a strong network bridging between the fiber and the matrix [12].

Although there are many different silane molecules available, the aminosilanes form the largest proportion of silanes employed in the composites industry [12]. The most common coupling agent is an aminosilane compound called γ-aminopropyltriethoxysilane (γ-APS), also known as APTES, which is the coupling agent in the studied sizing [16]. Usually sizings contain about 10 wt % of the coupling agent [29].

The composition of the sizing also consists of a number of multi-purpose components, such as a film former which holds the filaments together in a strand and protects the filaments from damage through fiber–fiber contact. Film formers are as closely compatible to the polymer matrix as possible. Epoxies, such as in this case, are very common film formers [12]. Usually sizings contain about 70–80 wt % of the film former [12,29].

Much less is known about the other chemicals in the sizing [12]. The sizing may also contain other compounds such as cationic or non-ionic lubricants, antistatic agents, emulsifiers, chopping aids, wetting agents or surfactants, and antioxidants [2,12,30]. Poly(propylene oxide) (PPO) or its co-polymer with poly(ethylene oxide) (PEO) is often used as a surfactant in sizings [2]. Polydimethylsiloxane (PDMS) is a common adhesion promoter, wetting agent, or surface tension reducer [2].

The exact composition of the sizing used in this study was not known to the authors, but based on technical details on the given R-glass fibers elsewhere [16], it is assumed that the sizing is based on the general characteristics described above. The results obtained are compatible with this assumption.

1.2. The Structure of the Sizing-Rich Composite Interphase

The structure of the sizing-rich composite interphase is very complex [12], as the sizing itself is heterogeneous and not uniform [12,31,32]. Furthermore, it has been observed by various researchers, that sizing is coated on fibers in "islands", "islets" or in patches, meaning that the fiber surface is only partially covered by the sizing, also giving some roughness to the surface [12,33–38]. Thomason and Dwight have concluded that epoxy-compatible sizings cover at least 90% of the glass fiber surface [39]. Mai et al. investigated APTES sizings using atomic force microscopy (AFM) and concluded that sized fibers are rougher than unsized fibers [38]. Also, similar conclusions were drawn by a few other researchers, including Turrión et al., who have shown that thickness of the sizing on the glass fibers varies from some nanometers up to a few hundred nanometers due to roughness [6,31,37].

With regards to the molecular structure of the interphase, APTES forms chemical covalent and physico-chemical hydrogen bonds and van der Waals interactions with the glass fibers and the amine epoxy [12,40]. The majority of APTES molecules which react with the glass surface can only form single Si–O–Si bonds with the glass due to steric limitations, while the vast majority of Si–O–Si bond formation in the silane interphase is due to polymerization—formation of the poly(siloxane) network [12]. A multilayer is formed on the glass fiber surfaces where the amino groups form intramolecular ring structures [32,41].

The concept of a composite interphase can be represented by a matrix polymer/poly(siloxane)/glass fiber model (shown in Figure 1) [5].

The siloxanes and poly(siloxanes) form covalent bonds with the glass fiber surface, resulting in a two-dimensional interface, the thickness of which is governed by the length of the chemical bonds, and is of an ångstrøm-scale (one tenth of a nanometer) [5].

The composite interphase is a gradient-type blend of the sizing compounds and the bulk matrix polymer, usually being about a micrometer in thickness [5,12,29,42,43]. It was observed, that an interfacial failure occurs at 0.5–4 nm from the glass surface in glass/γ-APS/epoxy interphase, indicating that the interphase region, rather than the two-dimensional interface is the weak link [5].

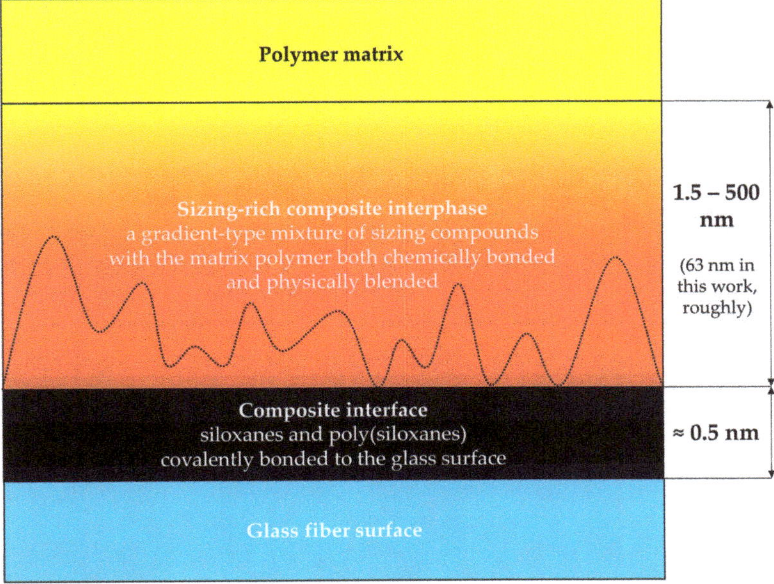

Figure 1. The concept of a polymer–siloxane–glass interphase, after [5]. The dotted line indicates that the sizing is rough [6,31,38].

1.3. The Aim of This Work

Composites take up water from their surroundings and may release some molecules into the surrounding water. Water uptake curves for composites are not straightforward to interpret since each constituent (matrix, fibers, sizing-rich interphase) interacts differently with the absorbed water. The mass uptake curve presents the combined effect of all these individual interactions.

Testing of water absorption is usually stopped when the composite material's water uptake has reached a maximum. The typical test procedures follow ASTM D5229 [44], where testing is stopped when two subsequent measurements do not differ by more than 0.5% [44]. However, when exposure to water is extended for longer periods, experiments performed in this work showed that degradation of the composite continues and additional mass gain and loss processes are involved. A similar observation was made by Perreux et al. [45], who studied immersion in water of 2.7 mm thick glass fiber epoxy composite plates for up to 10 years. They found that the weight gain of plates aged at 60 °C increased strongly after saturation. After about five years in water at 60 °C, a composite plate started to continuously lose mass with time [45]. This study will show that these effects can be related to the hydrolytic degradation of the fiber/matrix interphase.

The aim of this manuscript is to describe the degradation of the fiber/matrix interphase with special emphasis on the reaction kinetics.

2. Materials and Methods

2.1. Materials

Composite laminates were made with an amine-cured epoxy. The epoxy was prepared by mixing reagents Epikote Resin RIMR135™ (Hexion, Columbus, OH, USA) and amine based Epikure Curing Agent RIMH137™ (Hexion), stoichiometrically, in a ratio of 100:30 by weight. The mixture was degassed in a vacuum chamber for 30 min in order to remove bubbles. The density of the polymer (ρ_m) was 1.1 g/cm^3. Resin and hardener system consisted of the following compounds by composition: 0.63 wt % bisphenol A diglycidyl ether (DGEBA), 0.14 wt % 1,6-hexanediol diglycidyl ether (HDDGE), 0.14 wt % poly(oxypropylene)diamine (POPA) and 0.09 wt % isophorondiamine (IPDA) [46].

A typical glass fiber used for marine and oil and gas applications was selected: boron-free and fluorine-free high-strength, high-modulus 3B HiPer-Tex™ W2020 R-glass (3B-the fiberglass company, Hoeilaart, Belgium). Stitch-bonded mats were used. The average fiber diameter was 17 ± 2 μm [47,48]. The density of glass (ρ_f) was 2.54 g/cm^3 [47,48].

Composite laminates 50 mm thick were prepared via vacuum-assisted resin transfer molding (VARTM). Laminates were manufactured using the aforementioned fabrics and epoxy resin. The curing was performed at room temperature for 24 h, continued by post-curing in an air oven (Lehmkuhls Verksteder, Oslo, Norway) at 80 °C for 16 h. Full cure was achieved [46,49]. The composite laminates were cut into specimens with dimensions of 50 mm × 50 mm × 1.5 mm. The geometry of the samples and cross section of the fibers is shown in Figure 2. Two configurations C1 and C3 were cut, as shown in Figure 2. Configurations. C1 is representative of a typical composite where fibers are parallel to one of the long sides. The surface area of cut fibers with exposed cross sections is 50 mm × 1.5 mm. Configuration C3 was cut in a way that a maximum number of cut fibers were obtained having exposed cross sections (50 mm × 50 mm). The length of the fibers was just 1.5 mm. This unusual specimen was made to obtain maximum fiber exposure towards the water. The same specimens were also used to measure anisotropic diffusivity in a separate study [49]. The specified dimensions were achieved within 5% tolerance. The thickness was adjusted using a grinding and polishing machine Jean Wirtz PHOENIX 2000 (Jean Wirtz, Dusseldorf, Germany) and SiC discs (Struers, Cleveland, OH, USA; FEPA P500, grain size 30 μm).

Figure 2 also shows a micrograph of a surface with visible cross sections of cut fibers from a specimen with C3 configuration. The micrograph was taken with a confocal microscope InfiniteFocus G4 (Alicona, Graz, Austria).

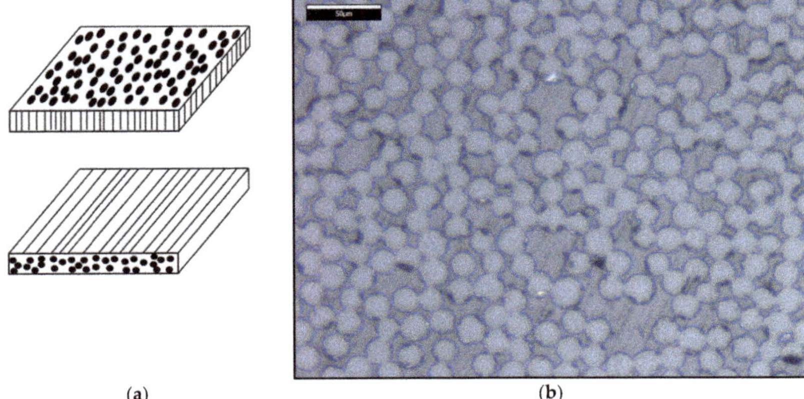

(a)   (b)

Figure 2. Glass fiber-reinforced epoxy composite plates: (**a**) sample configuration indicating alignment of the fibers in the plate: C3 at the top; C1 at the bottom; (**b**) micrograph of the largest face of the composite plate showing the cross section of the fibers at the surface.

Distilled water (resistivity 0.5–1.0 MΩ·cm) was used for conditioning of the composite samples. It was produced using the water purification system Aquatron A4000 (Cole-Parmer, Vernon Hills, IL, USA). The pH of the distilled water was 5.650 ± 0.010, being lower than neutral due to dissolved CO_2 from atmosphere in equilibrium.

2.2. Experimental Methods

2.2.1. Loss on Ignition

The loss on ignition (LOI) value of the fiber bundles was determined according to the standard practice ASTM D4963 [23]. This technique allows measurement of the weight loss of a sized glass sample. Since the weight loss is due to the burning off of the sizing, the method can be used to determine the amount of sizing on the fiber [12]. According to the LOI measurements, the sizing was 0.64 wt % of the sized fibers. The temperature during the LOI measurement was about 565 °C applied for about 5–5.5 h.

The obtained LOI is consistent with literature. LOI of most glass fiber reinforcement products is below 1.2 wt % [12]. For instance, Zinck and Gerard [50] also studied an APTES-based sizing which had a similar LOI value of 0.77 wt %.

2.2.2. Constituent Volume and Mass Fractions of the Composite

The fiber volume fraction of the composite was 59.5% and was determined using the burn-off test, after the ASTM Standard D3171 [51]. The void volume fraction of the composite was 0.44% and was measured by image analysis of optical microscope images, as was described elsewhere by Gagani et al. [49]. Fiber, matrix, interphase and voids volume fractions were 59.5%, 39.2%, 0.9% and 0.44%, respectively. The interphase volume fraction was obtained using the LOI value (0.64 wt %), the mass of sized glass fibers (about 5.6 g), the density of the interphase (1.1 g/cm^3) and the mass of the composite (about 7.2 g). Fiber, matrix and interphase mass fractions were 77.2%, 22.3% and 0.5%, respectively. The fiber surface area of one plate was about 0.5 m^2 on average. The composite interphase mass fraction (m_{f_i}) was calculated as:

$$m_{f_i} = \frac{\text{LOI} \cdot m_{\text{fibers}}}{m_{\text{comp}}} \tag{1}$$

where m_{fibers} is the mass of the sized fibers; m_{comp} is the mass of the composite.

2.2.3. Conditioning of Composite Plates

Water uptake and hygrothermal aging of the composite laminates was conducted using a batch system. A heated bath with distilled water (60 ± 1 °C) was used for conditioning the samples. Samples were weighed using analytical scales AG204 (±0.1 mg; Mettler Toledo, Columbus, OH, USA). Samples were conditioned for a period of about a year. Three parallels were performed.

2.2.4. Specific Surface Area of the Fibers Obtained by N_2 Sorption/Desorption and Brunauer–Emmett–Teller (BET) Theory

The specific surface area of the sized and unsized glass fibers was obtained via N_2 sorption and desorption. The method uses physical adsorption and desorption of gas molecules based on the Brunauer–Emmett–Teller (BET) theory [52]. The specific surface area was measured using QUADRASORB SI (Quantachrome Instruments, Boynton Beach, FL, USA) equipment. BET tests for specific surface area determination were performed according to the international standard ISO 9277:2010(E) [53]. The method is based on the determination of the amount of adsorptive gas molecules covering the external surface of the solid [53].

Since the sizing's surface is rough [12], the BET tests can provide the specific surface area. The BET theory explains the physical adsorption of gas molecules on a solid surface of a material, and it is the basis for the specific surface area determination.

Due to the roughness of the sizing on the fiber surface, the specific surface area of sized glass fibers measured with BET was 0.180 m²/g (see Figure 3), being higher than the specific surface area of unsized glass fibers of 0.09 m²/g (geometrical considerations as described in Section 2.1) or 0.084 m²/g using the BET method. For the unsized and sized glass fibers, the data with the BET model fit was with a determination coefficient R^2 of 0.968 and 0.994, respectively.

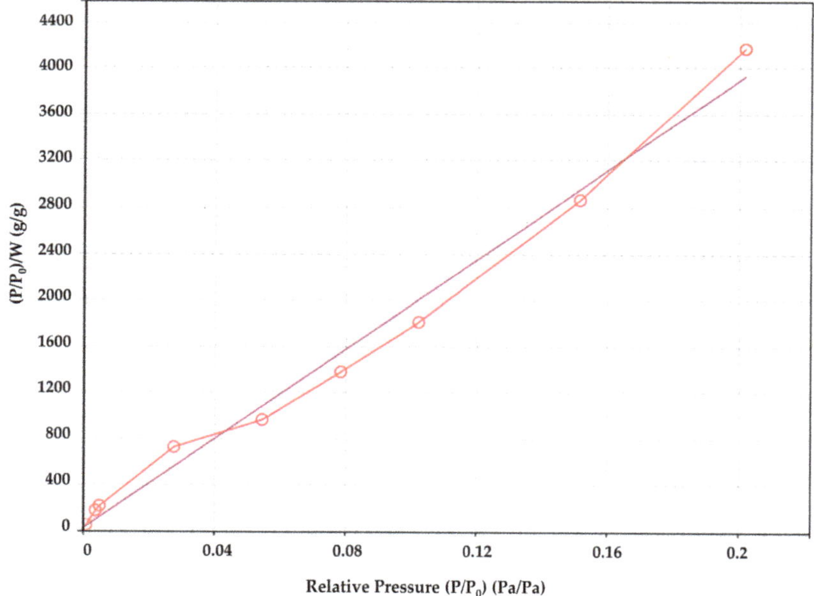

Figure 3. Brunauer–Emmett–Teller (BET) analysis of the specific surface area of the sized fibers.

3. Analytical Model

3.1. Mass Balance

When polymers take up water from the environment, their mass is affected by the water uptake itself, leaching and aging mechanisms such as hydrolysis, chain scission or oxidation [46,54]. For the studied epoxy, there is no significant mass loss due to chemical bond scission, since hydrolysis and chain scission are not occurring [46,55].

The combination of the phenomenological perspective and mass balance approach provide a useful tool for analyzing mass uptake/loss processes in composites during hygrothermal aging by breaking down a complex process into constituent-related processes. The processes that affect weight gain or loss of composites are summarized in Table 1.

Table 1. Summary of the processes during hygrothermal aging of composites that affect the mass balance.

Process	Sign	Reference
Water uptake of the polymer matrix	+	[49]
Water uptake by the composite interphase	+	[56]
Water uptake by the voids	+	[12,49,57]
Thermo-oxidation of the polymer matrix	+	[46]
Leaching from polymer matrix	−	[46]
Glass fiber dissolution	−	[28,47]
Sizing-rich interphase dissolution	−	This work

Gravimetric measurements determine the sample's mass over time during conditioning in water. The mass consists of the following terms:

$$m_{\text{gravimetric}}(t) = m_{\text{dry}} + m_{\text{water uptake}}(t) + m_{\text{oxidation}}(t) - m_{\text{leaching}}(t) \\ - m_{\text{glass dissolution}}(t) - m_{\text{interphase dissolution}}(t) \tag{2}$$

The dissolution of the interphase is then simply given by:

$$m_{\text{interphase dissolution}}(t) = m_{\text{dry}} + m_{\text{water uptake}}(t) + m_{\text{oxidation}}(t) - m_{\text{leaching}}(t) \\ - m_{\text{glass dissolution}}(t) - m_{\text{gravimetric}}(t) \tag{3}$$

The proposed model equation should be a phenomenologically full representation of the interaction between the composite material and the water environment. More details will now be given for each of the terms.

3.2. Water Uptake

The water uptake for composites includes three sub-processes: the uptake by the polymer matrix, by the interphase, and by the voids [49]. The glass fibers themselves do not absorb any water.

The water taken up by the polymer matrix at any point of time is limited by the diffusivity and the water saturation level [49,54]. The Fickian diffusion model can be used to model the water uptake by the polymer and the interphase [49]. In addition, the effect of voids being filled with water has to be considered [12,49]. The water content at saturation of the studied epoxy is 3.44 wt % if no voids are present [56]. Saturation has been defined as the moment when the difference in two consecutive water absorption measurements is lower than 0.5%, as defined by ASTM D5229 [44]. The composite's saturation water content M_∞ was determined to be 0.96 wt % [49]. It can be calculated by Equation (4) [49]:

$$M_\infty = \frac{M_\infty^m (v_{\text{m}} + v_{\text{i}})\rho_{\text{m}} + M_\infty^v v_{\text{v}} \rho_{\text{water}}}{v_{\text{f}}\rho_{\text{f}} + (v_{\text{m}} + v_{\text{i}})\rho_{\text{m}}} \tag{4}$$

where ρ_m is the matrix density, ρ_f is the fiber density, ρ_{water} is the water density, v_f is the fiber volume fraction, v_m is the matrix volume fraction, v_i is the interphase volume fraction, v_v is the void volume fraction ($v_f + v_m + v_i + v_v = 1$), M_∞^m is the matrix saturation water content (3.44 wt %) and M_∞^v is the void saturation water content (100 wt %). Fiber, matrix, interphase and voids volume fractions are 59.5%, 39.2%, 0.9% and 0.44%, respectively.

The sizing-rich interphase is assumed to have the same saturation water content as the epoxy matrix, since it contains about 70–80 wt % epoxy film-former [12,29,56]. Since the volume of the sizing is very small compared to the composite's volume any deviation from this assumption would have a minimal effect on the water uptake.

It is assumed here that the small voids will be completely filled with water $M_\infty^v = 1$, as was measured experimentally for the composite described here [49].

The water diffusivity of the studied epoxy polymer and the composites C1 and C3 in the thickness direction (with the fibers running transverse and parallel to the thickness direction for C1 and C3, respectively; see Figure 2) are systematized in Table 2 [49]. The higher diffusivity of the composite C3 is due to the fact that the diffusivity of the interphase in the direction parallel to the fibers is almost an order of magnitude higher than that of the polymer, after [49].

Table 2. Diffusivities in the through-the-thickness direction, after [49].

Specimen	D (mm²/h)
Epoxy	0.0068
C1	0.0051
C3	0.0210

The following equation links the mass uptake to diffusivity from solving the 1-D Fickian diffusion equation, as described by Crank [58]:

$$M(t) = M_\infty\left[1 - \left(\frac{8}{\pi^2}\right)\sum_{i=0}^{\infty} \frac{e^{-(2i+1)\left(\frac{\pi}{h}\right)^2 Dt}}{(2i+1)^2}\right] \tag{5}$$

By fitting the exact solution of the diffusion equation to an exponential function, the ASTM standard simplified equation is the following [44]:

$$M(t) = M_\infty\left[1 - e^{-7.3\left(\frac{Dt}{h^2}\right)^{0.75}}\right] \tag{6}$$

where $M(t)$ is the water content, M_∞ is the water saturation content, t is time, h is the thickness and D is the diffusivity in the thickness direction of the plate.

More details and 3-D Fickian model calculations can be found elsewhere [49]. 1-D and 3-D Fickian models gave the same result. Thus, for the sake of simplicity, the 1-D diffusion model for water uptake is used in this work.

Experimental gravimetric measurements and modeled water uptake curves using Equation (6) are shown in Figure 4 for a composite C3 with and without voids. It can be clearly seen that the absorption of water in the voids needs to be modeled to get a good fit with the experimental data.

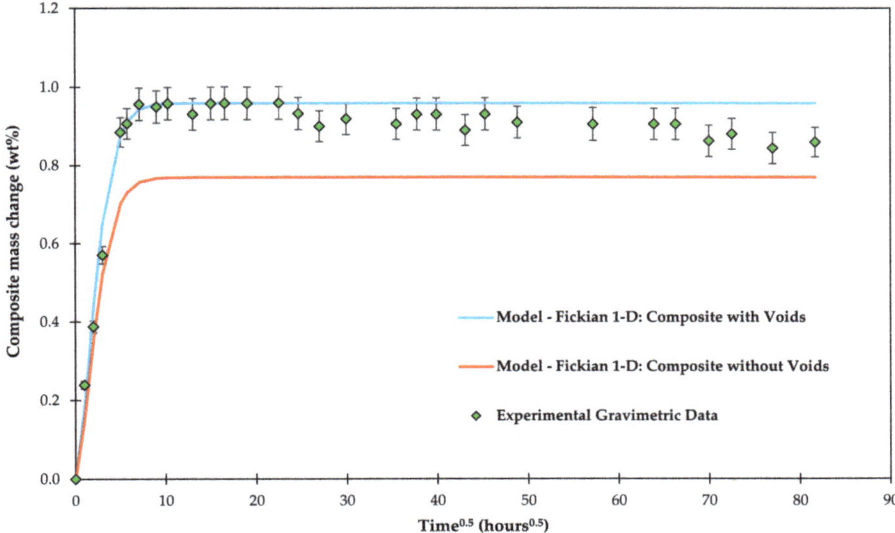

Figure 4. Experimental gravimetric measurements of composite C3 plates conditioned in water and modeled water uptake curves using the Fickian 1-D model.

3.3. Oxidation of the Epoxy Matrix

Photo-oxidation is not present as the material is not exposed to high-energy irradiation [46]. The effect of thermo-oxidation on mass gain due to water uptake is negligible. Thermo-oxidation for the studied epoxy polymer occurs via the carbonyl formation mechanism in the carbon-carbon backbone via nucleophilic radical attack, as is described elsewhere [46].

3.4. Leaching of Molecules out of the Epoxy Matrix

Water molecules can migrate into the epoxy polymer while at the same time small molecules may leach out of the matrix [59,60]. The leaching phenomenon may occur due to initially present additives, impurities, unreacted hardener or degradation products diffusing out of the epoxy network into the water environment, which is in contact with the polymer. Often leaching follows Fickian-type diffusion [61]. The driving force of this process is due to the difference in concentration of these chemicals inside the polymer, and in the surrounding aqueous environment.

Leaching was determined experimentally using HR-ICP-MS up to about 1100 h in another work for the same epoxy material as used for making the composites in this study [46]. Krauklis and Echtermeyer [46] found that for the studied epoxy polymer there was no leaching of hardener, whilst the leaching occurred of epoxy compounds and impurities, such as epichlorohydrin and inorganic compounds. Based on Fourier transform–near infrared (FT-NIR) spectra (reported in [46]) the leached amount after about 1100 h of conditioning was estimated to be at 54.74 wt % of the initial leachable compounds present in the material. This indicates that more than a half of the small molecules were leached out after the relatively short time of 1100 h. The initial leachable compound content $M^0_{leaching}$ was found to be 0.092 wt % (about 1.5 mg) defined as the mass loss due to leaching divided by the initial mass of the polymer (about 1.6 g).

The diffusivity of leached compounds through the epoxy polymer was determined according to 1-D Fickian diffusion [44,61]:

$$M_{\text{leaching}}(t) = M_{\text{leaching}}^0 \left[1 - e^{-7.3\left(\frac{D_{\text{leaching}}t}{h^2} \right)^{0.75}} \right] \tag{7}$$

The diffusivity was obtained by regression analysis of the data performing non-linear Generalized Reduced Gradient (GRG) algorithm, while minimizing the residual sum of squares. The leaching diffusivity D_{leaching} obtained in this study was 6.0×10^{-5} mm^2/h.

The leached-out compounds were experimentally measured with High-resolution inductively coupled plasma mass spectrometry (HR-ICP-MS) (data from [46]). The modeled leaching behavior from the matrix polymer is shown in Figure 5.

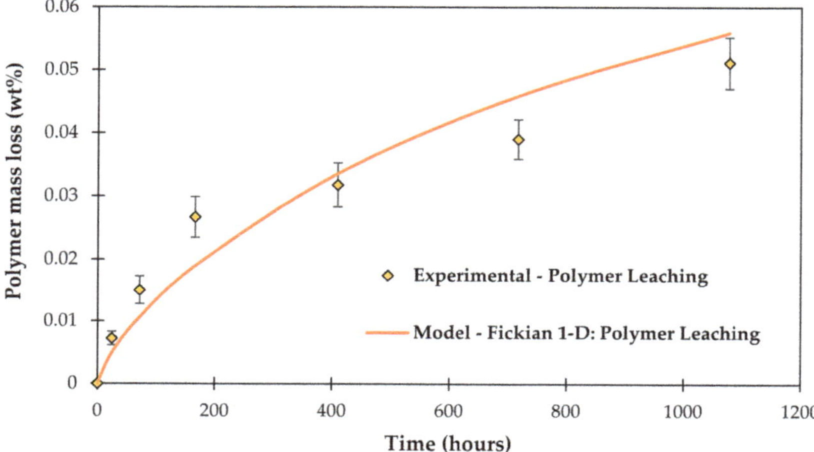

Figure 5. Polymer leaching determined experimentally with HR-ICP-MS, after [46], and modeled using Fickian 1-D model, after [61].

3.5. Glass Dissolution

Glass fibers slowly degrade in water environments via dissolution reactions resulting in a mass loss [47,62,63]. The degradation of glass fibers follows two distinct kinetic regions: short-term non-steady-state (Phase I) and long-term steady-state degradation (Phase II), as described in the dissolving cylinder zero-order kinetic (DCZOK) model for prediction of long-term dissolution of glass from both fiber bundles [47]. During Phase I, the degradation is complex and involves such processes as ion exchange, gel formation and dissolution. When Phase II is reached, the dissolution becomes dominant and the degradation follows zero-order reaction kinetics. For the studied R-glass, the transition from Phase I into Phase II occurs in about a week (166 h) at 60 °C and pH 5.65 [28,47]. Elements that are released during degradation of R-glass are Na, K, Ca, Mg, Fe, Al, Si and Cl [47]. The glass mass loss is the cumulative mass loss of all these ions [47]. Si contribution to the total mass loss of the studied R-glass is the largest (56.1 wt %) and seems to govern the dissolution process [47].

The rate of the dissolution depends on the apparent glass dissolution rate constant (K_0^*) and the glass surface area exposed to water (S) [28,47]. The glass surface area is proportional to the fiber radius. As the dissolution continues, the radius decreases linearly with time resulting in the mass loss deceleration; the DCZOK model accounts for this effect [47]. Rate constants at various environmental conditions (pH, temperature and stress), as well as more details about the model can be found in other works [28,47,63].

For a thin composite with fibers parallel to the short side through-thickness direction, such as in this work, the dissolution of glass, compared to the free fiber bundles with sizing (not embedded in the composite), is slowed down by 36.84% [28]. The differential mass loss equation for thin composites can be written as [28]:

$$\frac{\partial m}{\partial t} = K_0^* S(t) \tag{8}$$

The K_0^* includes the effects of diffusion and accumulation of the degradation products inside the composite, the protective effect of the sizing and the availability of water [28,47,63]. The time-dependent parameter is the fiber surface area $S(t)$.

Considering the two distinct phases of the degradation, the full DCZOK model in the integral form is the following, after [47]:

$$\begin{cases} t \leq t_{st} : \; m_{\text{dissolved}} = n\pi l \left(2r_0 K_0^{*\,I} t - \dfrac{K_0^{*\,I\,2}}{\rho_f} t^2 \right) \\ t > t_{st} : \; m_{\text{dissolved}} = m_{\text{dissolved}_{t_{st}}} + n\pi l \left(2r_{t_{st}} K_0^{*\,II}(t - t_{st}) - \dfrac{K_0^{*\,II\,2}}{\rho_f}(t - t_{st})^2 \right) \end{cases} \tag{9}$$

where n is the number of fibres (6450824); l is the length of fibres (1.5 mm); r_0 is the initial fiber radius (8.5 μm), and ρ_{glass} is the density of glass (2.54 g/cm^3); $K_0^{*\,I}$ and $K_0^{*\,II}$ are the apparent dissolution rate constants (g/m^2·s) for the short-term non-steady-state (Phase I) and long-term steady-state (Phase II) regions, respectively; $r_{t_{st}}$ (m) and $m_{\text{dissolved}_{t_{st}}}$ (g) are the fiber radius and lost mass after time t_{st} (s), when steady-state is reached (166 h [47,63]).

Using the composition of dissolving ions reported for the studied R-glass (Si contribution 56.1 wt %) [47], and the composite data after [28], $K_0^{*\,I}$ and $K_0^{*\,II}$ for the studied composite are 6.91×10^{-6} and 1.54×10^{-6} g/(m^2·h), respectively. The dissolution rate constants are systematized in Table 3. The glass mass loss was modeled using the DCZOK Equation (9) as shown in Figure 6. The glass mass loss is normalized by the composite plate's glass fiber surface area (about 0.5 m^2).

Table 3. Apparent glass dissolution rate constants.

Phase	K_0^* (g/(m^2·h))
Phase I	6.91×10^{-6}
Phase II	1.54×10^{-6}

These ions determined with HR-ICP-MS come from both glass material and the sizing-rich interphase. HR-ICP-MS can capture ions from interface and interphase (ionic products of the polysiloxane/siloxane hydrolysis), but ICP does not allow carbon detection due to CO_2 in the plasma, thus the organics from sizing-rich interphase are not captured. In other words, the predicted mass loss due to dissolution includes ions coming from the interface and interphase, but does not include organic compounds from the interphase. This is what makes the difference between the HR-ICP-MS determined mass loss and the gravimetric mass loss of the composite.

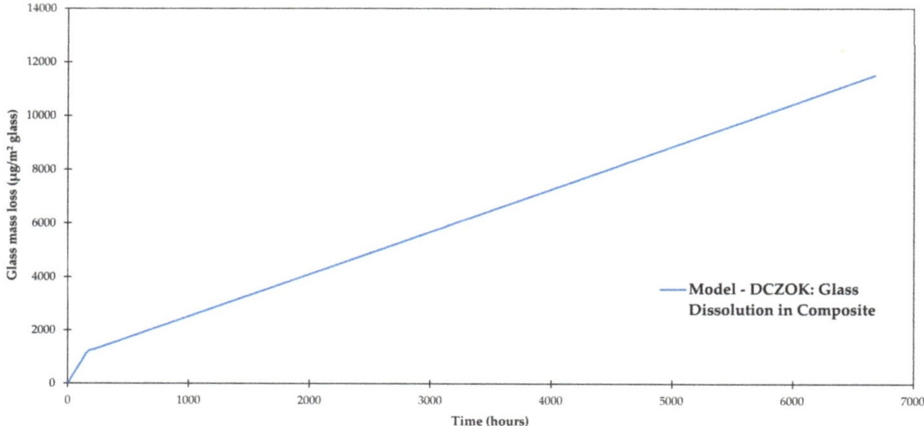

Figure 6. Glass-fiber dissolution modeled using dissolving cylinder zero-order kinetics (DCZOK) for the studied composite, after [28,47,63].

3.6. Interphase Dissolution

The aging of the sizing-rich composite interphase is the least understood constituent. The small amount of the interphase sizing compared to the composite bulk material makes analysis difficult. The proprietary nature of the sizing's composition allows only general evaluations. For typical sizing formulations, water interacting with the interphase may hydrate the Si–O–Si and Si–O–C bonds [5]. It was found that water molecules adsorbed in the epoxy matrix could migrate towards the sizing/glass fiber interface through the sizing, resulting in the dissolution/decomposition of the polysiloxane [30]. The reaction with water breaks strained Si–O–Si bonds and generates Si–OH sites [12]. Principle silane chemical bonding is reversible in the presence of water, thus the Si–O–Si bonds can be broken due to hydrolysis, as shown in Chemical Reaction (10) [12]:

$$\equiv \text{Si–O–Si} \equiv +\text{H}_2\text{O} \leftrightarrow 2 \equiv \text{Si–OH} \tag{10}$$

In this work, the sizing-rich interphase loss is modeled assuming a simple zero-order kinetic model.

4. Results and Discussion

The increase of the composite's mass with time within the first few hundred hours could be fairly well described by a standard diffusion approach, as shown in Figure 4. It was important to include the water uptake of the voids in the calculations. However, the diffusion approach would predict a constant mass over time once saturation has been reached (0.96 wt %). The data of C3 show a slight gradual drop in mass after saturation was reached, whereas the mass of C1 is clearly increasing, as shown in Figure 7.

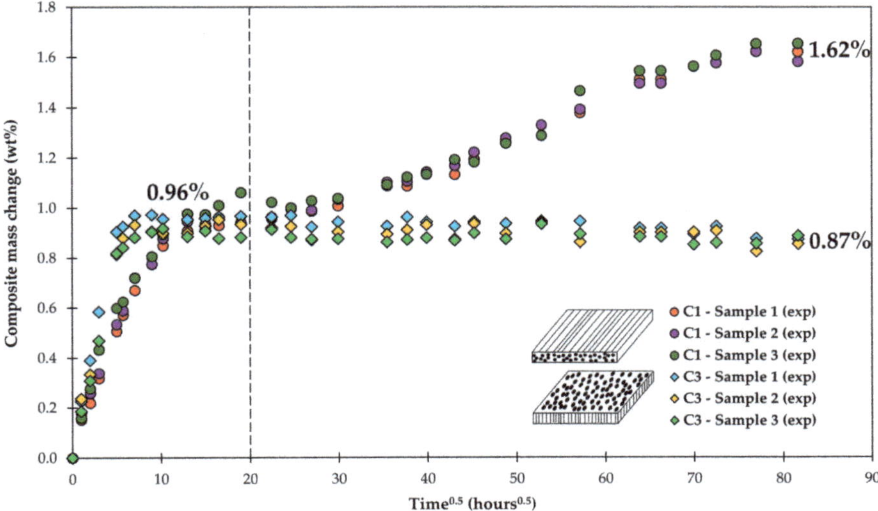

Figure 7. Long-term water uptake by composite laminates. Dashed line corresponds to a time when a test following standard practice ASTM D5229 would be stopped [44].

If the water uptake experiments are stopped as suggested by ASTM, then the long-term behavior is not captured. This observation is also consistent with the results of another study on long-term water uptake by composite plates [45]. The diverging behavior of water uptake by C1 and C3 composites can be observed starting only after about 20 $h^{0.5}$ (about 2 weeks), only after the saturation M_∞ (0.96 wt %) has been already achieved. The discussion on how the diverging behavior of C1 and C3 can be captured will follow.

4.1. Samples with Short Fibers C3

Firstly, the gravimetric behavior of C3 is addressed. As described above, a mass loss can be caused by leaching material out of the epoxy and by the glass fibers losing ions. If these effects are added to the mass vs. time curve a fairly good agreement with the experimental data is achieved, as shown in Figure 8. It could be argued that the agreement is sufficient within the experimental scatter. However, a closer look at the data can give some insight in the behavior of the sizing (interphase), although the evaluation is at the limit of what can be analyzed considering the scatter of the results.

Looking at Figure 8, a slightly better fit of the data can be obtained with a curve that has a higher mass loss with increasing time, i.e., is a bit steeper. This extra loss of material could be related to the disintegration of the interphase. The simplest approach is to model the mass loss of the interphase using the zero-order kinetics [64]:

$$\frac{\partial m_i}{\partial t} = K_i^0 S_i(t) \tag{11}$$

where m_i is the mass of the interphase, K_i^0 is the kinetic coefficient of the interphase dissolution and S_i is the surface area of the interphase. The solution of this equation for cylindrical fibers is given in Equation (9). For small mass changes and short times, the equation can be approximated by its first linear term with the sizing having a constant surface area S_i^0 to be:

$$m_i(t) = m_i^0 - K_i^0 S_i^0 t \tag{12}$$

The initial mass of the sizing m_i^0 (35.7 mg) was determined by the burn-off test to be 0.64 wt % of the sized fibers. Fitting the data in Figure 8 allows finding $K_i^0 S_i^0$, which basically describes the slightly steeper slope compared to the previous analysis based only on matrix and glass fiber dissolution. Using linear regression, as shown in Figure 9, The best fit for $K_i^0 S_i^0 = 1.80 \times 10^{-7}$ g/h.

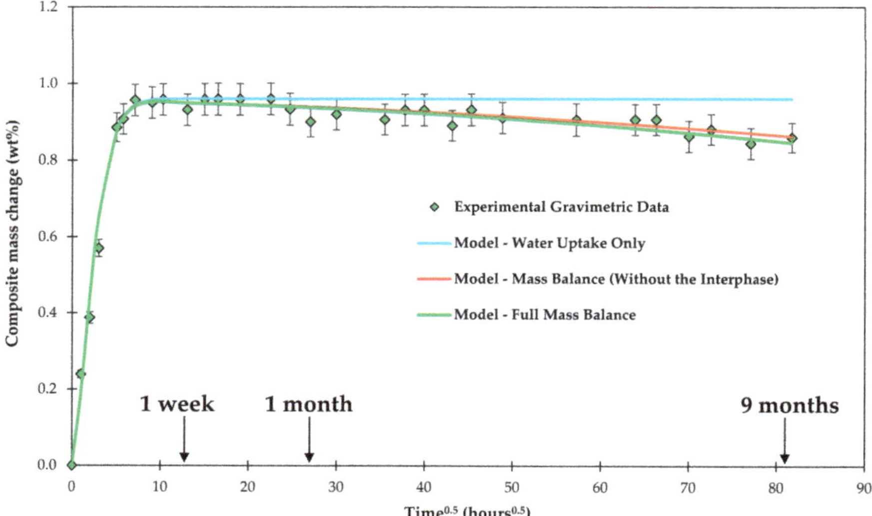

Figure 8. Experimental composite C3 plate mass change during the conditioning in water, shown over a square root of time. Water uptake and mass balance are modeled.

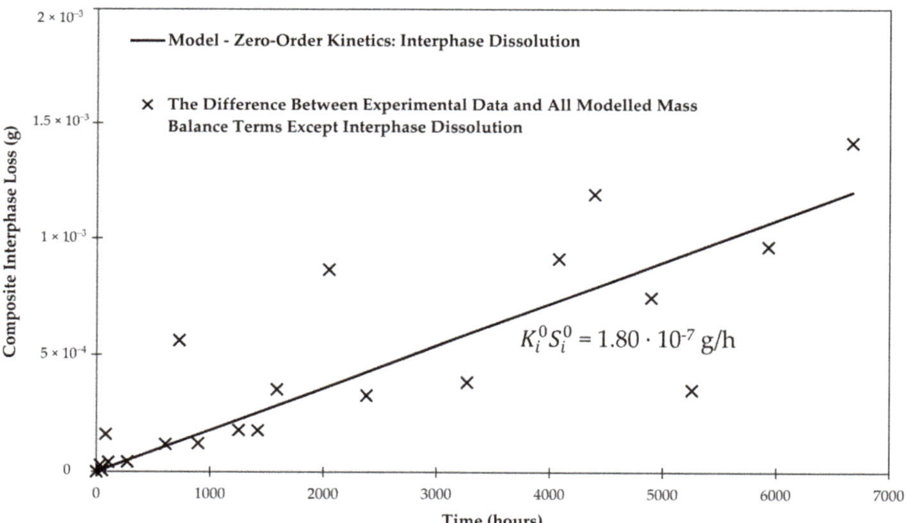

Figure 9. Linear regression of the difference between the experimental data and the all modelled terms except the interphase. The regressed line provides insight about the rate of the composite interphase dissolution in water.

Since dissolution is a surface reaction, a surface area of the sizing-rich interphase has to be obtained in order to determine the kinetics of dissolution. Unfortunately, we do not know the exact surface area of the sizing. Using the BET method, it was found that unsized fibers have a surface area of

0.084 m^2/g and sized fibers have a surface area of 0.180 m^2/g, roughly twice the value of the unsized fibers. As discussed in the introduction, the sizing is rough which creates a larger surface [6,31,37], but it also covers only parts of the fiber [12,32–37]. A typical sizing coverage of 90% of the glass fiber surface is assumed, after [39]. Furthermore, the sizing is bonded to the glass fiber on one side and the epoxy matrix on the other side, which does not create free surfaces at all. Based on the currently available information, the only possibility is to calculate K_i^0 for a number of plausible scenarios for the surface area S_i^0.

The thickness of the interphase is obtained from the volume of the interphase V_i taking geometry and known coverage (i.e., 0.9 or 1) into consideration. The volume of the interphase is known from LOI (0.64 wt %; 35.8 mg) and interphase density (1.1 g/cm^3), $V_i = 0.0325$ cm^3. The thickness of the interphase is then obtained as follows:

$$\delta_i = \frac{V_i}{\text{Coverage}\cdot S_{\text{glass}}} \tag{13}$$

where S_{glass} is the total glass fiber surface area in a composite plate (about 0.5 m^2). For 90% and 100% coverage, a mean interphase thickness is 72 and 65 nm, respectively.

Scenario 1. The minimum surface area S_i^0 would be just the cross-sectional area of the sizing exposed on the surface of the composite specimen. The fiber fraction was 59.5% and the area of one exposed surface of a C3 specimen was 50 mm × 50 mm. The surface area of fibers on both exposed surfaces is then 2975 mm^2. The radius of an individual fiber was 8.5 μm. Based on the burn-off method (LOI 0.64 wt %) and assuming extreme 100% coverage, the average sizing thickness was 65 nm. The ratio of exposed sizing cross sectional area to fiber cross sectional area is then 0.0149 and the exposed sizing area is 44.3 mm^2. In this scenario $K_i^0 = 4.06 \times 10^{-3}$ g/(m^2·h). The sizing would be dissolved along the axis of the fibers while the exposed cross section would remain constant until the sizing is completely dissolved. Equation (12) would accurately describe dissolution in this scenario. For these 1.5 mm-thick samples, the time to dissolve the sizing would be 22.7 years.

Scenario 2. The other extreme would be to argue that the epoxy is quickly saturated with water (after about 100 and 81 h for C1 and C3, respectively), The water can then attack and dissolve the sizing. In that case, the exposed area of the sizing would be much bigger. The BET method measured a specific surface area of sized fibers to be 0.180 m^2/g. Then, the total surface area of sized fibers (5.6 g fibers) in one plate is 1.01 m^2. Since the sizing covers only parts of the fiber, not all of this surface is from the sizing. But to obtain an outer bound K_i^0 can be calculated for this maximum surface area (assuming coverage of 100%). In this case using Equation (12), $K_i^0 = 1.78 \times 10^{-7}$ g/(m^2·h). The K_i^0 should be accurately determined by this equation for the relatively small area reduction during the measurement. However, the proper cylindrical Equation (9) taking the surface area reduction with time into account should be used to obtain the long-term dissolving of the sizing. The time to dissolve the sizing would be 30.5 years.

Scenario 3. Considering the descriptions of the literature about sizing, a typical sizing covers approximately 90% of the fiber [39]. In that case, the surface area of the sizing would be 0.91 m^2. Using the same approach of a cylindrical sizing exposed to water in the epoxy as described for Scenario 2 above the K_i^0 for this case would be 1.98×10^{-7} g/(m^2·h) and the time to dissolve the sizing would be 30.5 years.

The parameters of the three scenarios are systematized in Table 4.

Table 4. Systematized scenarios of the interphase dissolution kinetics.

Scenario	$K_i^0 S_i^0$ (g/h)	Sizing Coverage (%)	δ_i (nm)	S_i^0 (m^2)	K_i^0 (g/(m^2·h))	Time to Total Dissolution (years)
Scenario 1	1.80×10^{-7}	100	65	4.43×10^{-5}	4.06×10^{-3}	22.7
Scenario 2	1.80×10^{-7}	100	65	1.01	1.78×10^{-7}	30.5
Scenario 3	1.80×10^{-7}	90, after [39]	72	0.91	1.98×10^{-7}	30.5

The mass loss due to long-term gravimetric behavior of composite C3 could be successfully modeled, because the C3 samples did not have a significant accumulation of the degradation products. The C3 plates have a short fiber length (1.5 mm). Once the matrix is saturated with water, the water can attack and degrade the interphase. Any reduction products can be quickly transported along the interphase to the surface of the sample and will be absorbed by the surrounding water.

4.2. Samples with Long Fibers C1

The C1 samples showed a mass increase with time, see Figure 7, an additional 0.66 wt % of water was taken up after 6673 h of conditioning. Since C1 and C3 samples were made from the same laminate, just cut in a different direction, the change in behavior must be related to the sample's geometry. Compared to the C3 samples the C1 samples have much longer fibers and subsequently much longer fiber matrix interphases (1.5 mm vs. 50 mm).

The matrix of both sample types absorbs water in roughly the same period (see Table 2). The water will attack the interphase between fibers and matrix in the same way. But, it is believed that degradation products (of fibers and interphase) cannot easily move along the interphase and escape into the surrounding water at the composite's surface. Instead, the weakening of the interphase causes the formation of flaws. The degradation products and water can accumulate in these flaws. Thus, the mass of the composite does not decrease with time as for samples C3, but the mass of C1 samples increases with time. Figure 10 shows schematically what such a flaw could look like. Figure 11A shows that such flaws are, indeed, observed in the samples.

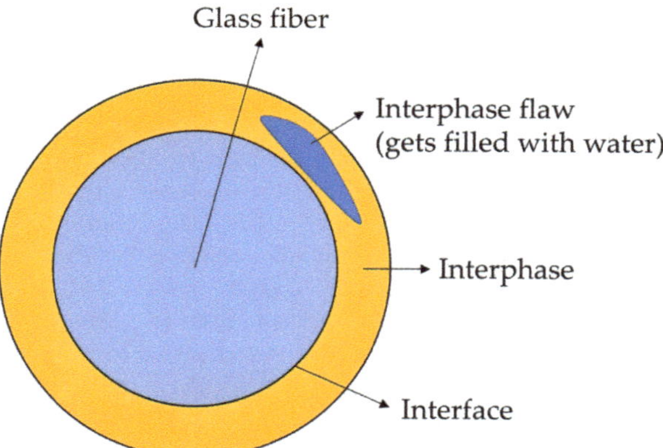

Figure 10. Interphase flaw is formed and is filled with water.

Since the laminate absorbed another 0.66 wt % of water, it is possible to estimate the size of flaws needed to accommodate this amount of water. The initial mass of the C1 plate was about 7.36 g. Water in the interphase flaws should thus weigh 48.6 mg, taking up volume of 4.86×10^{-8} m^3. Assuming for the moment that all fibers have evenly distributed flaws, the following calculations can be made. Dividing volume necessary to accommodate the extra water by the amount of fibers in a composite C1 plate (193525) and the length of a fiber (50 mm), the cross-sectional area of a water-filled interphase flaw around one fiber is found to be 5.02×10^{-12} m^2. The radius of the glass fiber is 8.5 µm, thus the cross-sectional area of the fiber is 2.27×10^{-10} m^2. By combining cross-sectional areas of the interphase flaw and the fiber, and deducting the radius of the fiber, an average thickness of a water-filled interphase flaw of 93.5 nm is obtained.

In reality, not all interphase flaws are the same size and not all fiber/matrix interphases are damaged equally, as shown in Figure 11. The weakest links will fail first. Once cracks are formed, stresses are released and more complicated processes follow. However, it is interesting that the first fiber matrix debondings, as shown in Figure 11A, have dimensions similar to the calculated value of 93.5 nm. Fiber/matrix debondings shown in Figure 11A range from about a 100 nm to a few microns, as was observed experimentally using microscopy after 6673 h of conditioning. The thickness also matches debonding dimensions observed elsewhere for the same composite [65,66].

Three damage mechanisms were observed in the micrographs:

- Fiber/matrix debondings, shown in Figure 11A.
- Matrix transverse cracks, shown in Figure 11B. These cracks seem to be inside the bundle. This location may be also a result of the weakening of the fiber/matrix interphase, which was covered in point 1.
- Splitting along the fibers, shown in Figure 11C.

Fiber/matrix debonding appears to be the first failure mechanism, caused by hydrolysis of the interphase. This failure mechanism is described by the observations made for the C3 samples in Section 4.1 When these failure mechanisms accumulate, creating a weakened local region, they can easily combine into a longer "matrix crack" due to a release of curing, thermal and swelling stresses, resulting in a crack formation. The reason for the observed splitting along fibers is less clear. It could be related to the matrix cracks, but it could also be caused by the fibers used for stitching the reinforcing mat. All these flaws (cracks) create volume that can be filled with water and increases the mass of the composite.

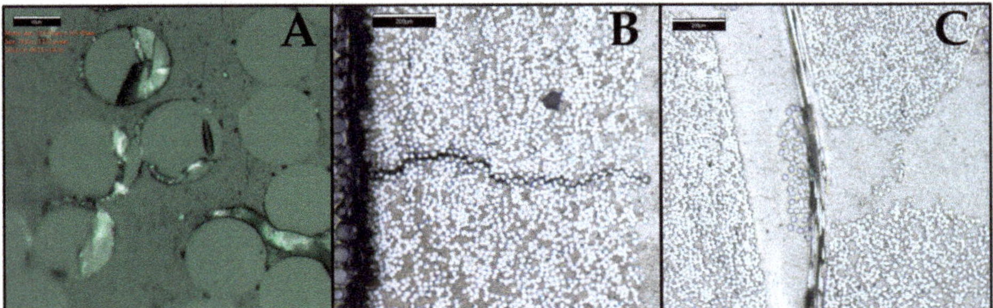

Figure 11. Micrograph of a composite sample exposed to water for 6673 h at 60 °C. The micrograph indicates the (**A**) fiber/matrix debondings; (**B**) matrix transverse cracks; (**C**) splitting along the fibers.

Perreux, Choqueuse and Davies [45] investigated long-term water uptake by 2.7 mm-thick composite plates. The plate was made with an anhydride-based curing agent while this study looked at an epoxy laminate made with an amine-based curing agent. They observed that after ASTM saturation was achieved, there was still a significant continuous mass gain up to about 5 years of conditioning in water at 60 °C. After this point, an abrupt and continuous mass loss occurred for the following 5 years until the measurements were stopped. The data is schematically shown in Figure 12.

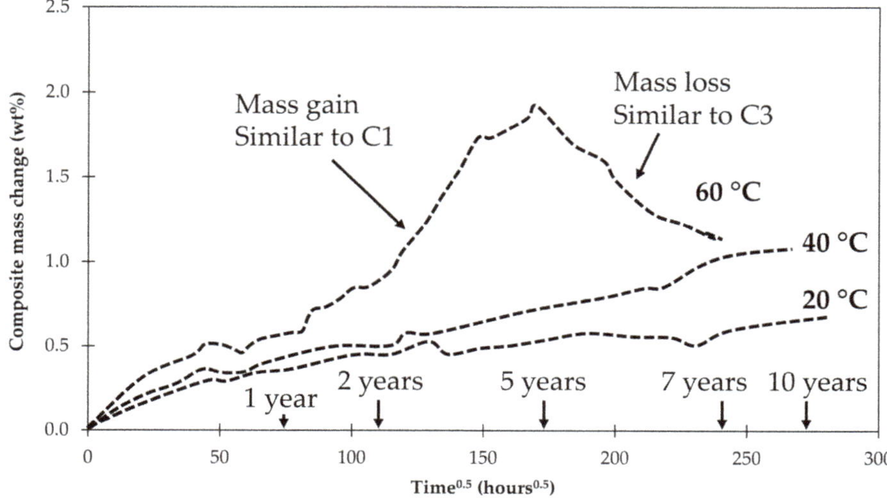

Figure 12. Schematic representation of the long-term water uptake at 20, 40 and 60 °C observed by Perreux, Choqueuse and Davies for the 2.7 mm thick composite plates [45].

The results seem to be a combination of what was found for samples C1 and C3 investigated here. An explanation for the behavior observed by Perreux et al. [45] may be given by the findings of this work. Initially flaws form in the composite interphase that is filled with water, resulting in a mass gain as found in C1 samples. At one point, so many flaws have accumulated that an open interpenetrating network with access to the surface of the laminate has formed. This network allows the degradation products and previously absorbed water to diffuse out, creating a mass loss similar to samples C3 (since fiber lengths in C3 were so small the interpenetrating network was present from the beginning). Since the observed mass reduction happened gradually, this means that the network of flaws and cracks is gradually being connected to the sample's surface. The mass drop was not observed for tests carried at lower temperatures. In that case all processes are slower and the samples only increased their mass, indicating the formation of flaws and cracks. But a network of the cracks reaching the surface was not created yet. It should be noted that the matrix of anhydride-based epoxies studied by Perreux et al. [45] is also prone to hydrolysis, so the hydrolysis in their samples may have affected the matrix and the fiber–matrix interphase.

4.3. General Aspects

For the composite laminates studied here, about 3.5 wt % of the interphase was dissolved in a year's time. The expected total dissolution for the geometry of the C3 sample would occur between 22.7 and 30.5 years, according to the three scenarios at 60 °C. At lower temperatures, the processes would be significantly slower, because diffusivities and dissolution rates follow Arrhenius-type temperature dependence [63,67]. Activation energies of these individual processes differ. Thus, it is likely not a straightforward Arrhenius-type influence on the process rate as a whole (summary mass uptake or loss).

The degradation time (22.7 to 30.5 years) should be independent of sample geometry and should be applicable once the matrix has reached saturation. For thick composite laminates the fiber–matrix interphase may only degrade in the surface region, because the matrix in the inside may remain dry. Degradation may also be stopped or slowed down by an accumulation of reaction product, if the degradation reaction is reversible, such as reaction (a). The mass uptake data obtained here showed a slight slowdown of the reaction after 9 months, close to the point when experiments were stopped, see

Figure 7. But it is unclear whether the data really flatten out. The test results from Perreux et al. [45] run over 10 years indicate that the degradation continues all the time.

Damage caused by the hydrolytic aging of the sizing-rich composite interphase very likely leads to a decrease in interfacial strength. For instance, Gagani et al. [68] and Rocha et al. [42] have reported the composite interphase-related deterioration of the mechanical properties due to aging in water. It is likely that the formation of the interphase flaws described in this work is the mechanistic origin of the interfacial strength deterioration of composites.

The authors think that studying the effect of seawater on the hydrolysis of the interphase would be useful, since the composite marine structures are most often used in the seawater environments. The dissolution in seawater conditions is expected to occur slower than in distilled water due to the presence of silica (dissolved from sand and other minerals). The reason for an expected aging rate slowdown in seawater is that the degradation products are already present in the surrounding environment, thus decreasing the driving force—a concentration gradient.

The length of glass fibers should not affect the molecular structure or morphology of the interphase per se. However, it should be added that what is affected by the fiber length is the path (or length) that the hydrolytic degradation products have to travel in order to escape the composite material and diffuse out into the surrounding water. It was shown in this work that water interaction with composites with very short interphase leads to mass loss, whereas for a typical composite an interpenetrating flaw network takes a relatively long time to form in order for degradation products to leave the composite. This leads to another aspect that needs to be studied in more detail: a diffusion of degradation products through the interphase. It is important to understand whether there is a diffusion-controlled aspect.

This paper covers hydrolysis of the composite interphase, but the same approach should be applicable for all other environmental agents and solvents (in general, solvolysis).

5. Conclusions

Glass fiber composites absorb water with time and the mass of the composites increase subsequently. When measuring diffusivity and saturation level of water according to ASTM D5229 [44] testing is stopped when the mass increase with time stops, i.e., it is reaching a plateau, in this case at about 200 h. However, continuing the tests exposing the laminates to water for longer, the mass of the composite increases again, measured up to 9 months. This additional water uptake was found to be due to the hydrolytic degradation of the sizing-rich fiber matrix interphase. Due to water-induced dissolution interphase flaws being formed which developed further into matrix cracks. The internal volume created by the flaws and cracks can be filled with water leading to the observed mass increase. The microscopically measured size of the flaws matches the order of magnitude of the volume required for obtaining the measured additional mass increase.

The hydrolytic degradation of the fiber matrix interphase could be investigated directly by cutting non-typical specimens from a thick composite laminate. The test specimens were 50 mm × 50 mm × 1.5 mm coupons where all the fibers were running parallel to the short edge. This created specimens with a short fiber–matrix interface length and the interphases being connected to the large sample's surface. When these specimens were conditioned in water, their mass increased during the first 200 h as the typical specimens described above. Continuing the test for longer times leads, however, to a mass loss. For these specimens, the flaws created by the fiber matrix interphase hydrolysis were open towards the surface of the test specimen, since the interphase length (and fiber length) was so short, 1.5 mm. The reaction products of the hydrolysis could migrate into the surrounding water bath leading to a mass drop. This mass loss allowed the product of the dissolution rate constant and the surface area of the interphase $K_i^0 S_i^0$ to be determined. The small specimens tested here would degrade the entire interphase within 22 to 30 years at 60 °C. The calculation is based on a full mechanistic mass balance approach considering all the composite's constituents: water uptake and leaching of the matrix, dissolution of the glass fibers, and dissolution of the composite interphase. These processes were modeled using a combination of Fickian diffusion and zero-order kinetics.

Based on long-term test data from the literature tested for close to 10 years, it seems that composites will initially absorb extra water in the flaws and cracks created by interphase hydrolysis. Eventually these cracks will create a network that is connected to the surface of the composite laminate. When this network is formed reaction products can leave the laminate and the mass will be reduced.

The possible strength degradation due to the flaws in the fiber matrix interface forming within 22 to 30 years (for the tested type of epoxy laminate) in saturated laminates should be taken into account in designs for long lifetimes.

Author Contributions: Conceptualization, A.E.K. and A.T.E.; Methodology, A.E.K.; Formal Analysis, A.E.K.; Investigation, A.E.K. and A.I.G.; Resources, A.E.K., A.I.G. and A.T.E.; Data Curation, A.E.K., A.I.G. and A.T.E.; Writing—Original Draft Preparation, A.E.K.; Writing—Review and Editing, A.E.K. and A.T.E.; Validation, A.E.K.; Visualization, A.E.K.; Supervision, A.T.E.; Project Administration, A.T.E.; Funding Acquisition, A.T.E.

Funding: This research was funded by The Research Council of Norway (Project 245606/E30 in the Petromaks 2 programme).

Acknowledgments: This work is part of the DNV GL led Joint Industry Project "Affordable Composites" with 19 industrial partners and the Norwegian University of Science and Technology (NTNU). The authors would like to express their thanks for the financial support from The Research Council of Norway (Project 245606/E30 in the Petromaks 2 programme). The authors are thankful to Erik Sæter, Valentina Stepanova, Susana Villa Gonzalez and Julie Asmussen. Andrey is especially thankful to Oksana V. Golubova.

Conflicts of Interest: The authors declare no conflict of interest.

Abbreviations

ρ_f	Density of the glass fibers (g/m^3)
ρ_m	Density of the matrix polymer (g/m^3)
ρ_i	Density of the sizing-rich composite interphase (g/m^3)
ρ_{water}	Density of the water (g/m^3)
h	Thickness of a material plate (m)
v_f	Volume fraction of the fibers (m^3/m^3)
v_m	Volume fraction of the matrix polymer (m^3/m^3)
v_i	Volume fraction of the composite interphase (m^3/m^3)
v_v	Volume fraction of the voids (m^3/m^3)
$M(t)$	Time-dependent water content of the composite (wt %)
M_∞	Saturation water content of the composite (wt %)
$M^m(t)$	Time-dependent water content of the matrix polymer (wt %)
M_∞^m	Saturation water content of the matrix polymer (wt %)
M_∞^v	Saturation water content of the voids (wt %)
D	Through-thickness water diffusivity of the material (mm^2/h)
$M_{leaching}(t)$	Time-dependent content of leached compounds from the polymer (wt %)
$M_{leaching}^0$	Initial leachable compound content in the polymer (wt %)
$D_{leaching}$	Through-thickness leachable compound diffusivity of the material (mm^2/h)
$r(t)$	Time-dependent fiber radius (m)
r_0	Initial fiber radius (m)
$r_{t_{st}}$	Fiber radius when the steady-state dissolution is reached (m)
K_0	Glass dissolution rate constant (g/(m^2·s))
K_0^*	Apparent glass dissolution rate constant (g/(m^2·s))
$K_0^{*\,I}$	Apparent glass dissolution rate constant (non-steady-state; Phase I) (g/(m^2·s))
$K_0^{*\,II}$	Apparent glass dissolution rate constant (steady-state; Phase II) (g/(m^2·s))
t_{st}	Time when long-term steady-state is reached (s)
n	Number of fibers (–)
l	Length of fibers and the interphase (m)

$S(t)$	Time-dependent glass fiber surface area (m^2)
S_0	Initial glass fiber surface area (m^2)
t	Time (s)
$m; m_{\text{dissolved}}$	Glass mass loss due to dissolution (g)
$m_{\text{dissolved}_{t_{\text{st}}}}$	Dissolved glass mass when the steady-state is reached (g)
ξ_{sizing}	Protective effect of sizing against glass dissolution (–)
n_{order}	Order of the water availability term (–)
$S_i(t)$	Time-dependent surface area of the composite interphase (m^2)
S_{i0}	Initial surface area of the composite interphase (m^2)
S_i^{specific}	Specific surface area of the composite interphase (m^2)
K_i^0	Zero-order rate constant of the composite interphase dissolution (g/(m^2·s))
$m_i(t)$	Time-dependent mass of the composite interphase (g)
m_{i0}	Initial mass of the composite interphase (g)
GF	Glass fiber
GFRP	Glass fiber-reinforced polymer; same as glass fiber-reinforced composite
DCZOK	Dissolving cylinder zero-order kinetic (model)
DGEBA	Bisphenol A diglycidyl ether
HDDGE	1,6-Hexanediol diglycidyl ether
POPA	Poly(oxypropylene)diamine
IPDA	Isophorondiamine
R-glass	"Reinforcement" glass
FRP	Fiber-reinforced polymer, same as fiber-reinforced composite
HR-ICP-MS	High-resolution inductively coupled plasma mass spectrometry\
VARTM	Vacuum-assisted resin transfer molding
BET	Brunauer–Emmett–Teller theory
LOI	Loss on ignition
γ-APS APTES	γ-aminopropyltriethoxysilane
PPO	Poly(propylene oxide)
PEO	Poly(ethylene oxide)
PDMS	Polydimethylsiloxane

References

1. Berg, J.; Jones, F.R. The role of sizing resins, coupling agents and their blends on the formation of the interphase in glass fiber composites. *Compos. Part A* **1998**, *29*, 1261–1272. [CrossRef]
2. Feih, S.; Wei, J.; Kingshott, P.; Sørensen, B.F. The influence of fiber sizing on the strength and fracture toughness of glass fiber composites. *Compos. Part A* **2005**, *36*, 245–255. [CrossRef]
3. Dai, Z.; Shi, F.; Zhang, B.; Li, M.; Zhang, Z. Effect of sizing on carbon fiber surface properties and fibers/epoxy interfacial adhesion. *Appl. Surf. Sci.* **2011**, *257*, 6980–6985. [CrossRef]
4. Yuan, X.; Zhu, B.; Cai, X.; Liu, J.; Qiao, K.; Yu, J. Optimization of interfacial properties of carbon fiber/epoxy composites via a modified polyacrylate emulsion sizing. *Appl. Surf. Sci.* **2017**, *401*, 414–423. [CrossRef]
5. DiBenedetto, A.T. Tailoring of interfaces in glass fiber reinforced polymer composites: A review. *Mater. Sci. Eng. A* **2001**, *302*, 74–82. [CrossRef]
6. Plonka, R.; Mäder, E.; Gao, S.L.; Bellmann, C.; Dutschk, V.; Zhandarov, S. Adhesion of epoxy/glass fiber composites influenced by aging effects on sizings. *Compos. Part A* **2004**, *35*, 1207–1216. [CrossRef]
7. Grabovac, I.; Whittaker, D. Application of bonded composites in the repair of ships structures—A 15-year service experience. *Compos. Part A* **2009**, *40*, 1381–1398. [CrossRef]
8. McGeorge, D.; Echtermeyer, A.T.; Leong, K.H.; Melve, B.; Robinson, M.; Fischer, K.P. Repair of floating offshore units using bonded fibre composite materials. *Compos. Part A* **2009**, *40*, 1364–1380. [CrossRef]
9. Gustafson, C.-G.; Echtermeyer, A. Long-term properties of carbon fibre composite tethers. *Int. J. Fatigue* **2006**, *28*, 1353–1362. [CrossRef]
10. Salama, M.M.; Stjern, G.; Storhaug, T.; Spencer, B.; Echtermeyer, A. The first offshore field installation for a composite riser joint. OTC-14018-MS. In Proceedings of the Offshore Technology Conference, Houston, TX, USA, 6–9 May 2002. [CrossRef]

11. Echtermeyer, A.T.; Gagani, A.I.; Krauklis, A.E.; Mazan, T. Multiscale modelling of environmental degradation—First steps. In *Durability of Composites in a Marine Environment 2. Solid Mechanics and Its Applications*; Davies, P., Rajapakse, Y.D.S., Eds.; Springer: Cham, Switzerland, 2018; Volume 245, pp. 135–149. ISBN 978-3-319-65145-3.

12. Thomason, J.L. *Glass Fiber Sizings: A Review of the Scientific Literature*; James L Thomason: Middletown, DE, USA, 2012; ISBN 978-0-9573814-1-4.

13. Weitsman, Y. Coupled damage and moisture-transport in fiber-reinforced, polymeric composites. *Int. J. Solids Struct.* **1987**, *23*, 1003–1025. [CrossRef]

14. Weitsman, Y.J.; Elahi, M. Effects of fluids on the deformation, strength and durability of polymeric composites—An overview. *Mech. Time-Depend. Mater.* **2000**, *4*, 107–126. [CrossRef]

15. Roy, S. Moisture-induced degradation. In *Long-Term Durability of Polymeric Matrix Composites*; Pochiraju, V.K., Tandon, P.G., Schoppner, A.G., Eds.; Springer: Boston, MA, USA, 2012; pp. 181–236. ISBN 978-1-4419-9307-6.

16. Peters, L. Influence of glass fibre sizing and storage conditions on composite properties. In *Durability of Composites in a Marine Environment 2. Solid Mechanics and Its Applications*; Davies, P., Rajapakse, Y.D.S., Eds.; Springer: Cham, Switzerland, 2018; Volume 245, pp. 19–31. ISBN 978-3-319-65145-3.

17. Culler, S.R.; Ishida, H.; Koenig, J.L. *Hydrothermal Stability of γ-Aminopropyltriethoxysilane Coupling Agent on Ground Silicon Powder and E-Glass Fibers*; Technical Report; Department of Macromolecular Science: Cleveland, OH, USA, 1983.

18. Wang, D.; Jones, F.R.; Denison, P. TOF SIMS and XPS study of the interaction of hydrolysed γ-aminopropyltriethoxysilane with E-glass surfaces. *J. Adhes. Sci. Technol.* **1992**, *6*, 79–98. [CrossRef]

19. Wang, D.; Jones, F.R.; Denison, P. Surface analytical study of the interaction between γ-amino propyl triethoxysilane and E-glass surface. Part I Time-of-flight secondary ion mass spectrometry. *J. Mater. Sci.* **1992**, *27*, 36–48. [CrossRef]

20. Wang, D.; Jones, F.R. Surface analytical study of the interaction between γ-amino propyl triethoxysilane and E-glass surface. Part II X-ray photoelectron spectroscopy. *J. Mater. Sci.* **1993**, *28*, 2481–2488. [CrossRef]

21. Wang, M.; Xu, X.; Ji, J.; Yang, Y.; Shen, J.; Ye, M. The hygrothermal aging process and mechanism of the novolac epoxy resin. *Compos. Part B* **2016**, *107*, 1–8. [CrossRef]

22. Halpin, J.C. *Effects of Environmental Factors on Composite Materials*; Technical Report AFML-TR-67–423; Air Force Materials Laboratory: Dayton, OH, USA, 1969.

23. *ASTM D4963/D4963M-2011 Standard Test Method for Ignition Loss of Glass Strands and Fabrics*; ASTM: West Conshohocken, PA, USA, 2011.

24. Loewenstein, K.L. *Glass Science and Technology (Book 6), The Manufacturing Technology of Continuous Glass Fibres*; Elsevier: Amsterdam, The Netherlands, 1993; ISBN 978-0444893468.

25. Thomason, J.L.; Adzima, L.J. Sizing up the interphase: An insider's guide to the science of sizing. *Compos. Part A* **2001**, *32*, 313–321. [CrossRef]

26. Plueddemann, E.P. *Silane Coupling Agents*, 2nd ed.; Plenum Press: New York, NY, USA, 1991; ISBN 978-0-306-43473-0.

27. Emadipour, H.; Chiang, P.; Koenig, J.L. Interfacial strength studies of fibre-reinforced composites. *Res. Mech.* **1982**, *5*, 165–176.

28. Krauklis, A.E.; Echtermeyer, A.T. Dissolving cylinder zero-order kinetic model for predicting hygrothermal aging of glass fibre bundles and fibre-reinforced composites. In Proceedings of the 4th International Glass Fibre Symposium, Aachen, Germany, 29–31 October 2018; pp. 66–72, ISBN 978-3-95886-249-4.

29. Joliff, Y.; Belec, L.; Chailan, J.-F. Impact of the interphases on the durability of a composite in humid environment—A short review. In Proceedings of the 20th International Conference on Composite Structures ICCS20, Paris, France, 4–7 September 2017.

30. Zhuang, R.-C.; Burghardt, T.; Mäder, E. Study on interfacial adhesion strength of single glass fiber/polypropylene model composites by altering the nature of the surface of sized glass fibers. *Compos. Sci. Technol.* **2010**, *70*, 1523–1529. [CrossRef]

31. Wolff, V.; Perwuelz, A.; El Achari, A.; Caze, C.; Carlier, E. Determination of surface heterogeneity by contact angle measurements on glass fibres coated with different sizings. *J. Mater. Sci.* **1999**, *34*, 3821–3829. [CrossRef]

32. Ishida, H.; Koenig, J.L. An investigation of the coupling agent/matrix interface of fiberglass reinforced plastics by fourier transform infrared spectroscopy. *Polym. Phys. B* **1979**, *17*, 615–626. [CrossRef]

33. Watson, H.; Mikkola, P.J.; Matisons, J.G.; Rosenholm, J.B. Deposition characteristics of ureido silane ethanol solutions onto E-glass fibres. *Colloids Surf. A* **2000**, *161*, 183–192. [CrossRef]

34. Feresenbet, E.; Raghavan, D.; Holmes, G.A. Influence of silane coupling agent composition on the surface characterization of fiber and on fiber-matrix interfacial shear strength. *J. Adhes.* **2003**, *79*, 643–665. [CrossRef]

35. Fagerholm, H.M.; Lindsjö, C.; Rosenholm, J.B.; Rökman, K. Physical characterization of E-glass fibres treated with alkylphenylpoly(oxyethylene)alcohol. *Colloids Surf.* **1992**, *69*, 79–86. [CrossRef]

36. Thomason, J.L.; Dwight, D.W. The use of XPS for characterization of glass fibre coatings. *Compos. Part A* **1999**, *30*, 1401–1413. [CrossRef]

37. Turrión, S.G.; Olmos, D.; González-Benito, J. Complementary characterization by fluorescence and AFM of polyaminosiloxane glass fibers coatings. *Polym. Test.* **2005**, *24*, 301–308. [CrossRef]

38. Mai, K.; Mäder, E.; Mühle, M. Interphase characterization in composites with new non-destructive methods. *Compos. Part A* **1998**, *29*, 1111–1119. [CrossRef]

39. Thomason, J.L.; Dwight, D.W. XPS analysis of the coverage and composition of coatings on glass fibers. *J. Adhes. Sci. Technol.* **2000**, *14*, 745–764. [CrossRef]

40. Wang, D.; Jones, F.R. TOF SIMS and XPS study of the interaction of silanized E-glass with epoxy resin. *J. Mater. Sci.* **1993**, *28*, 1396–1408. [CrossRef]

41. Chiang, C.H.; Ishida, H.; Koenig, J.L. The structure of aminopropyltriethoxysilane on glass surfaces. *J. Colloid Interface Sci.* **1980**, *74*, 396–404. [CrossRef]

42. Rocha, I.B.C.M.; Raijmaekers, S.; Nijssen, R.P.L.; van der Meer, F.P.; Sluys, L.J. Hygrothermal ageing behaviour of a glass/epoxy composite used in wind turbine blades. *J. Compos. Struct.* **2017**, *174*, 110–122. [CrossRef]

43. Kim, J.K.; Sham, M.L.; Wu, J. Nanoscale characterization of interphase in silane treated glass fibre composites. *Compos. Part A* **2001**, *32*, 607–618. [CrossRef]

44. *ASTM D5229/D5229M-14 Standard Test Method for Moisture Absorption Properties and Equilibrium Conditioning of Polymer Matrix Composite Materials*; ASTM International: West Conshohocken, PA, USA, 2014.

45. Perreux, D.; Choqueuse, D.; Davies, P. Anomalies in moisture absorption of glass fibre reinforced epoxy tubes. *Compos. Part A* **2002**, *33*, 147–154. [CrossRef]

46. Krauklis, A.E.; Echtermeyer, A.T. Mechanism of yellowing: carbonyl formation during hygrothermal aging in a common amine epoxy. *Polymers* **2018**, *10*, 1017. [CrossRef] [PubMed]

47. Krauklis, A.E.; Echtermeyer, A.T. Long-term dissolution of glass fibers in water described by dissolving cylinder zero-order kinetic model: Mass loss and radius reduction. *Open Chem.* **2018**, *16*, 1189–1199. [CrossRef]

48. *3B Fibreglass Technical Data Sheet*; HiPer-Tex W2020 Rovings: Belgium, Brussel, 2012.

49. Gagani, A.I.; Fan, Y.; Muliana, A.H.; Echtermeyer, A.T. Micromechanical modeling of anisotropic water diffusion in glass fiber epoxy reinforced composites. *J. Compos. Mater.* **2017**, *52*, 2321–2335. [CrossRef]

50. Zinck, P.; Gerard, J.F. On the hybrid character of glass fibres surface networks. *J. Mater. Sci.* **2005**, *40*, 2759–2760. [CrossRef]

51. *ASTM D3171/D3171-15 Standard Test Methods for Constituent Content of Composite Materials*; ASTM International: West Conshohocken, PA, USA, 2015.

52. Brunauer, S.; Emmett, P.H.; Teller, E. Adsorption of gases in multimolecular layers. *J. Am. Chem. Soc.* **1938**, *60*, 309–319. [CrossRef]

53. *International Standard ISO 9277:2010(E) Determination of the Specific Surface Area of Solids by Gas Adsorption—BET Method*; ISO: Berlin, Germany, 2010.

54. Popineau, S.; Rondeau-Mouro, C.; Sulpice-Gaillet, C.; Shanahan, M.E.R. Free/bound water absorption in an epoxy adhesive. *Polymer* **2005**, *46*, 10733–10740. [CrossRef]

55. Krauklis, A.E.; Gagani, A.I.; Echtermeyer, A.T. Hygrothermal aging of amine epoxy: reversible static and fatigue properties. *Open Eng.* **2018**, *8*, 447–454. [CrossRef]

56. Krauklis, A.E.; Gagani, A.I.; Echtermeyer, A.T. Near-Infrared Spectroscopic Method For Monitoring Water Content In Epoxy Resins And Fiber-Reinforced Composites. *Materials* **2018**, *11*, 586. [CrossRef]

57. Thomason, J.L. The interface region in glass-fibre-reinforced epoxy resin composites: 2. Water absorption, voids and the interface. *Composites* **1995**, *26*, 477–485. [CrossRef]

58. Crank, J. *The Mathematics of Diffusion*, 2nd ed.; Clarendon Press: Oxford, UK, 1975; ISBN 978-0-19-853411-6.

59. Maggana, C.; Pissis, P. Water sorption and diffusion studies in an epoxy resin system. *J. Polym. Sci. Part B* **1999**, *37*, 1165–1182. [CrossRef]

60. Toscano, A.; Pitarresi, G.; Scafidi, M.; Di Filippo, M.; Spadaro, G.; Alessi, S. Water diffusion and swelling stresses in highly crosslinked epoxy matrices. *Polym. Degrad. Stab.* **2016**, *133*, 255–263. [CrossRef]

61. Bruchet, A.; Elyasmino, N.; Decottignies, V.; Noyon, N. Leaching of bisphenol A and F from new and old epoxy coatings: Laboratory and field studies. *Water Sci. Technol.* **2014**, *14*, 383–389. [CrossRef]

62. Schutte, C.L. Environmental durability of glass-fiber composites. *Mater. Sci. Eng. R Rep.* **1994**, *13*, 265–323. [CrossRef]

63. Krauklis, A.E.; Gagani, A.I.; Vegere, K.; Kalnina, I.; Klavins, M.; Echtermeyer, A.T. Dissolution kinetics of R-glass fibres: Influence of water acidity, temperature and stress corrosion. *Fibers* **2019**, *7*, 22. [CrossRef]

64. Khawam, A.; Flanagan, D.R. Solid-state kinetic models: basics and mathematical fundamentals. *J. Phys. Chem. B* **2006**, *110*, 17315–17328. [CrossRef] [PubMed]

65. Gagani, A.I.; Mialon, E.P.V.; Echtermeyer, A.T. Immersed interlaminar fatigue of glass fiber epoxy composites using the I-beam method. *Int. J. Fatigue* **2019**, *119*, 302–310. [CrossRef]

66. Rocha, I.B.C.M.; van der Meer, F.P.; Raijmaekers, S.; Lahuerta, F.; Nijssen, R.P.L.; Mikkelsen, L.P.; Sluys, L.J. A combined experimental/numerical investigation on hygrothermal aging of fiber-reinforced composites. *Eur. J. Mech. Sol.* **2019**, *73*, 407–419. [CrossRef]

67. Bonniau, P.; Bunsell, A.R. Water absorption by glass fibre reinforced epoxy resin. In *Composite Structures*; Marshall, I.H., Ed.; Springer: Dordrecht, The Netherlands, 1981; pp. 92–105. ISBN 978-94-009-8122-5.

68. Gagani, A.I.; Krauklis, A.E.; Sæter, E.; Vedvik, N.P.; Echtermeyer, A.T. A novel method for testing and determining ILSS for marine and offshore composites. *Comp. Struct.* **2019**, *220*, 431–440. [CrossRef]

Article

The Effect of Texturing of the Surface of Concrete Substrate on the Pull-Off Strength of Epoxy Resin Coating

Kamil Krzywiński and Łukasz Sadowski *

Faculty of Civil Engineering, Wroclaw University of Science and Technology, Wybrzeże Wyspiańskiego 27, 50-370 Wrocław, Poland; kamil.krzywinski@pwr.edu.pl
* Correspondence: lukasz.sadowski@pwr.edu.pl; Tel.: +48-71-320-37-42

Received: 30 November 2018; Accepted: 18 February 2019; Published: 21 February 2019

Abstract: This paper describes a study conducted to evaluate the effect of texturing of the surface of concrete substrate on the pull-off strength (f_b) of epoxy resin coating. The paper investigates a total of seventeen types of textures: after grooving, imprinting, patch grabbing and brushing. The texture of the surface of the concrete substrate was prepared during the first 15 min after pouring fresh concrete into molds. The epoxy resin coating was laid after 28 days on hardened concrete substrates. To investigate the pull-off strength of the epoxy resin coating to the concrete substrate, the pull-off method was used. The results were compared with the results obtained for a sample prepared by grinding, normative minimal pull-off strength values and the values declared by the manufacturer. During this study twelve out of fifteen tested samples achieved a pull-off strength higher than 1.50 MPa. It was found that one of the imprinting texturing methods was especially beneficial.

Keywords: concrete substrate; texturing; adhesion; cohesion; epoxy resin; coating; pull-off strength

---

1. Introduction

The construction industry is growing intensively, especially in the field of large area floors. This is mainly due to the growth of the production and transport industries and the fact that larger areas are being adapted for storage. To ensure a suitable floor for more significant loads, epoxy resins are mainly used as concrete surface coatings [1,2]. This usually allows a satisfactory pull-off strength (f_b) of coatings to be obtained [3]. The destruction of epoxy resin coating occurs during freezing and thawing processes [4], chemical compound aggression [5], erosion and corrosion [6,7], or resistance against thermal shock [8]. As pointed out by Garcia and de Brito [9], the major advantages of epoxy resin coatings are: its high chemical and mechanical resistance, being easy to clean, and its watertightness. Epoxy resin coatings are used to enhance the durability properties of a floor [10,11]. They increase the service life of a floor and decrease its failure [12]. They may also be used as preventive repair [13] and surface protection.

Before the application of epoxy resin coating, the manufacturer recommends preparing the surface of the concrete substrate using sandblasting or grinding, cleaning it, and then using a bonding agent to achieve a guaranteed pull-off strength (usually 2.0 MPa). Usually, grinding is the most effective method of mechanically treating the surface of the concrete substrate before the application of the epoxy resin coating. This was quantified by using 3D roughness parameters of the concrete substrate [14]. However, these steps are labor-intensive and expensive. Moreover, there is a higher probability of failure during these steps. In the authors' opinion there is a need to search for a way to avoid these steps during the construction process of epoxy resin coating floors. One recent example of such a procedure is to modify the composition of the epoxy resin coating using nanosilica [15], carbon nanotubes [16], glass powder [17], polymers [18] or diacrylate monomers [19]. These attempts were successful in increasing the pull-off strength of the epoxy resin coating to the concrete substrate. However, these kinds of modifications usually have a negative effect on the other important properties of epoxy resins, e.g., mechanical strength and viscosity. Thus, there is a need to find another way to improve the pull-off strength of the epoxy resin coatings [20]. It seems sensible to search for a proper method of treating the concrete substrate before the application of the epoxy resin coating.

Texturing of the surface of the concrete substrate is a promising method [21]. It may be especially effective when carried out on the surface of the fresh and liquid material of the substrate. For example, He et al. [22] used brushing to obtain a satisfactory coarseness of the surface of the concrete substrate. Alternatively, Mirmoghtadaei et al. [23] textured the surface of the concrete substrate using grooving and brushing in order to obtain a satisfactory coarseness. According to the authors, there have been no attempts to investigate the effect of texturing of the surface of concrete substrate on the pull-off strength of the epoxy resin coating. The following questions also remain unanswered: Which method of texturing is the most useful? Will imprinting or patch grabbing be effective?

When considering the above, the purpose of this manuscript is to evaluate the effect of texturing of the surface of concrete substrate on the pull-off strength of the epoxy resin coating. The paper also aims to ensure a proper pull-off strength of the epoxy resin coating by treating the surface of the concrete substrate using different texturing methods. For this purpose, four different texturing methods were used: grooving, imprinting, patch grabbing and brushing. Pull-off strength results were compared with the value obtained for the grinded concrete substrate surface, the value of the normative minimal pull-off strength and the value declared by the manufacturer.

2. Materials and Methods

2.1. Concrete Substrate

The concrete substrate samples were prepared in wooden forms measuring $150 \times 150 \times 40$ mm^3. To decrease the friction between the wood and sample, internal walls were covered with oil. The 40 mm thick substrate was prepared using a ready mix concrete of class C16/20 (Baumit, Wrocław, Poland). This composition consists of Type 1 Portland cement, quartz aggregate, limestone powder, sand with a grain size of 0–4 mm, and other additives. In this study the weight water-binder ratio of the ready-mix was 0.1 and the mixing time was 3 min. This kind of concrete is commonly used in civil engineering as a concrete substrate for epoxy resin coatings.

2.2. Texturing of the Surface of the Concrete Substrate

The surface of a freshly laid and liquid concrete mixture of the substrate was textured in four basic ways: grooving, imprinting, patch grabbing and brushing. The names and the descriptions of applied texture methods have been summarized in Table 1. All of these four methods are also shown in Figures 1–4.

Finally, for the grooving, widely available nets (No. 1) or grids (No. 2) with different size holes and thicknesses were used (Figure 1).

The imprinting was created using small cross spacers with thicknesses of 2, 4 and 6 mm. Moreover, two pattern types were designed for the plastic spacers: type "+" with crosses in regular spacing (Nos. 3, 5, 7), and type "x" with an additional rotated (45°) cross in the middle of four regular spacers (Nos. 4, 6, 8). The general view of the applied imprinting method is shown in Figure 2.

The patch grabbed textures were created with the use of a flooring trowel with different square sizes (Nos. 10–12), and also with a 2 mm rotated corrugated nylon tube along the concrete substrate surface (No. 9). The third designed surface method is shown in Figure 3.

The brushing was prepared with the use of brushes of three hardness levels: delicate bristles (No. 13), medium delicate bristles (No. 14.) and wire bristles (No. 15). After brushing, a texture depth between 2 and 3 mm has been obtained. The fourth method with medium delicate bristles is presented in Figure 4.

Table 1. The names and the descriptions of applied textured methods.

Name of the Texturing Method	Sample Number	Description of Applied Texturing Method
Grooving	1	Grooving using a building net to produce a series of grooves with a depth of 1 ± 0.5 mm and a gap from 7 to 9 mm between each other
	2	Grooving using a painting grid to produce a series of grooves with a depth of 8 ± 0.5 mm and with a gap from 9 to 11 mm between each other
Imprinting	3	Imprinting using small cross plastic spacers type "+" with a thickness of 2 mm in regular spacing of 34 mm
	4	Imprinting using small cross plastic spacers type "x" with a thickness of 2 mm in regular spacing of 24 and 34 mm
	5	Imprinting using small cross plastic spacers type "+" with a thickness of 4 mm in regular spacing of 34 mm
	6	Imprinting using small cross plastic spacers type "x" with thickness of 4 mm in regular spacing of 24 and 34 mm
	7	Imprinting using small cross plastic spacers type "+" with a thickness of 6 mm in regular spacing of 34 mm
	8	Imprinting using small cross plastic spacers type "x" with thickness of 6 mm in regular spacing of 24 and 34 mm
Patch grabbing	9	Patch grabbing with a rotated corrugated nylon tube with the diameter of 2 mm
	10	Patch grabbing with a flooring trowel with size of 4×4 mm^2
	11	Patch grabbing with a flooring trowel with size of 6×6 mm^2
	12	Patch grabbing with a flooring trowel with size of 12×12 mm^2
Brushing	13	Brushing with a painting brush with width of 150 mm
	14	Brushing with a wire brush with width of 190 mm
	15	Brushing with a normal brush with width of 140 mm

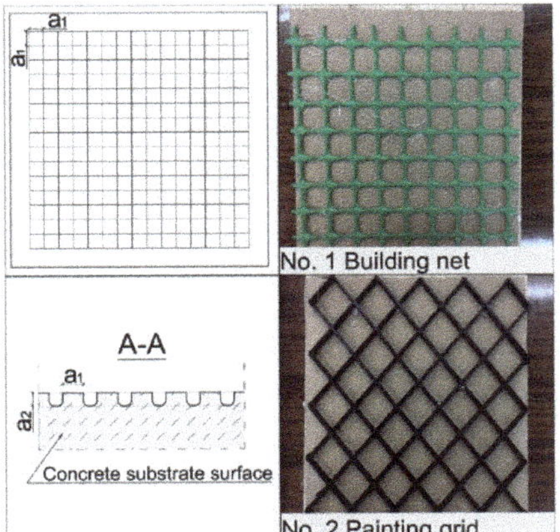

Figure 1. The applied grooving method for texturing the surface of the concrete substrate (dimensions: $a_1 = 8, 10$ mm; $a_2 = 1.0, 8$ mm).

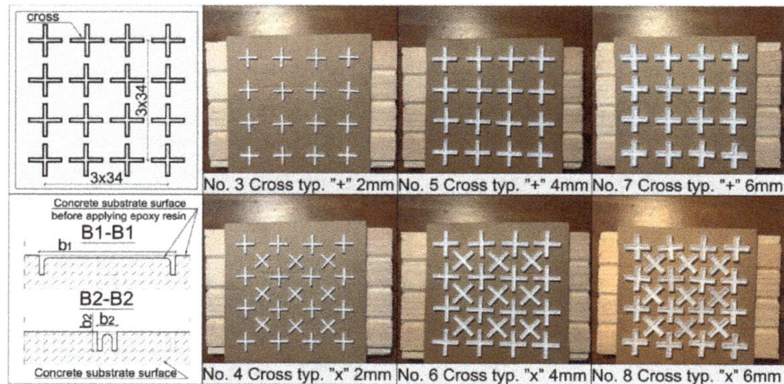

Figure 2. The applied imprinting method for texturing the surface of the concrete substrate (dimensions: b_1 = 20, 28 mm; b_2 = 2, 4, 6 mm).

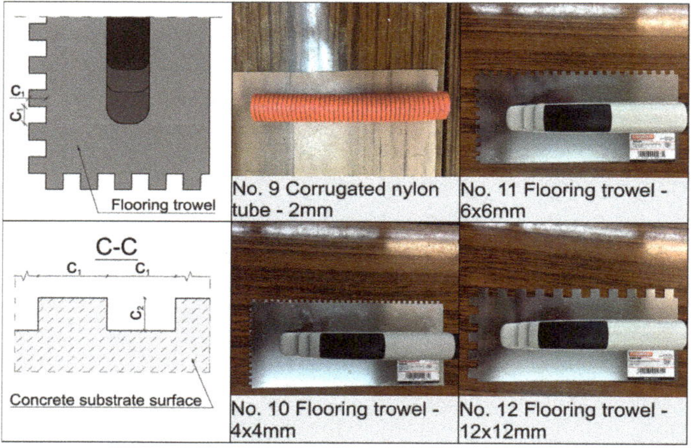

Figure 3. The applied patch grabbing method for texturing the surface of the concrete substrate (dimensions: c_1 = 2, 4, 6, 12 mm; c_2 = 1, 4, 6 mm).

Figure 4. The applied brushing method for texturing the surface of the concrete substrate.

The exemplary optical views of the surfaces of the concrete substrates after 28 days of maturation are presented in Figure 5.

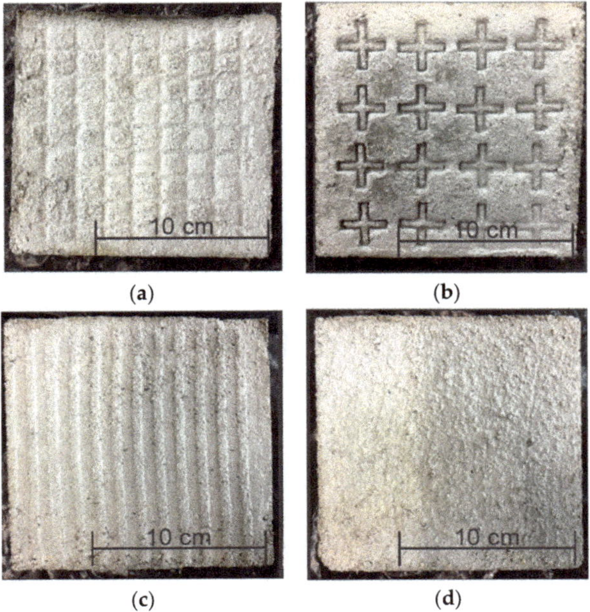

Figure 5. Exemplary optical views of the surfaces of the concrete substrates after 28 days of maturation: (**a**) grooved; (**b**) imprinted; (**c**) patch grabbed; (**d**) brushed.

The manufacturer of the epoxy resin recommended that the concrete surface be treated in two steps: grinding and applying a bonding agent. These actions allow the value of the pull-off strength higher than 2.0 MPa to be obtained after the epoxy resin coating is hardened for seven days. In this study, the concrete substrates were only textured without grinding and applying a bonding agent. For comparative purposes, one sample surface was grinded manually using a grinding stone with ceramic abrasive grain (No. X) in order to compare the obtained values with the textured forms and the pull-off strength declared by the manufacturer (f_b > 2.0 MPa). The exemplary optical view of the surface of the concrete substrate after grinding has been presented in Figure 6. On this surface no bonding agent has been applied. These values were also compared to the minimum value of the pull-off strength required by the standard EN 1542 [24] (f_b > 1.5 MPa).

Figure 6. Exemplary optical view of the surface of the concrete substrate after grinding.

2.3. Epoxy Resin Coating

Commercially available epoxy resin (StoPox BB OS, Sto-ispo Sp. z o.o., Wrocław, Poland) was prepared from two components. The first, which is a base, is an epoxy resin based on bisphenol (Component A). The second component is a hardener based on aliphatic polyamines (Component B). The weight ratio of A:B is 100:25. A plastic knife was used to mix the two components together for 3 min in order to obtain a uniform consistency (Figure 7).

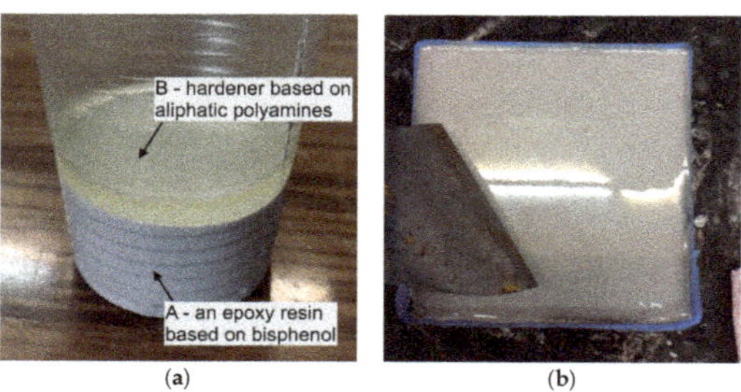

Figure 7. Epoxy resin coating: (**a**) Components–A & B; (**b**) sample after pouring fresh epoxy resin.

30 min after mixing, the material is suitable to be applied at 20 °C. The viscosity of the epoxy resin after mixing with the hardener was in a range between 1400 to 2300 MPa. The epoxy resin was aged in a controlled laboratory environment at the temperature of 20 ± 2 °C and a relative humidity less than 65%. The epoxy resin obtains enough strength for the pull-off strength tests seven days after being poured.

2.4. Pull-Off Strength Tests

The automatic adhesion tester (DY-216, Proceq, Schwerzenbach, Switzerland) was used for the pull-off strength test according to ASTM D4541 [25]. This method has recently become very popular in assessing the pull-off strength of polymer modified coatings [26]. For each sample one specimen was tested in three places. During the test, the load on the fixture was increased in a manner that was as smooth and continuous as possible. The rate of the load was 0.05 MPa/s in order for failure to occur or so that the maximum stress was reached in about 100 s or less. After obtaining the test results, the type of failure and concrete substrate detached thickness were also analyzed (Figure 8).

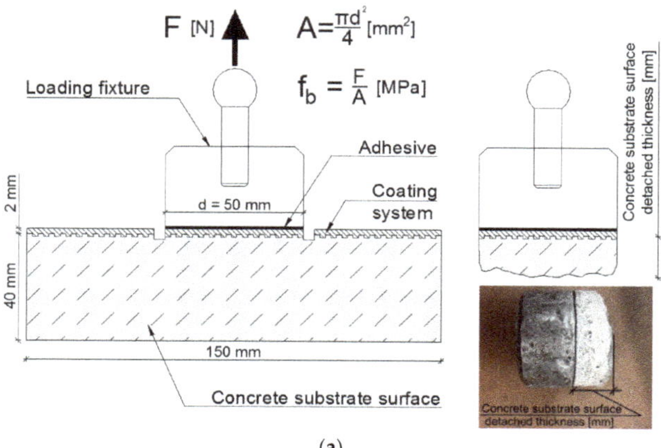

Figure 8. *Cont.*

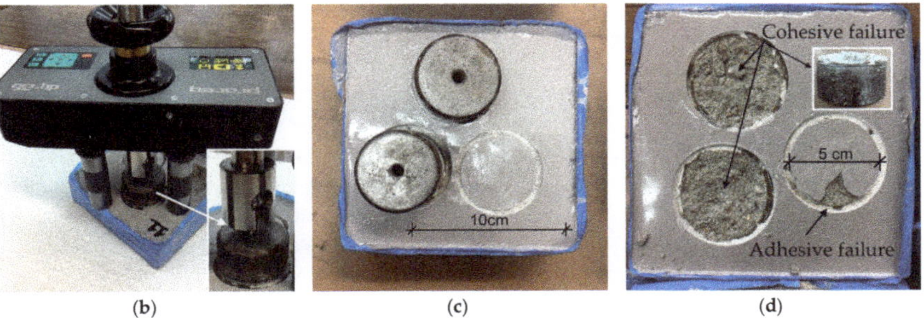

Figure 8. Pull-off method: (**a**) scheme of the pull-off strength test; (**b**) Proceq dy-216 measuring the pull-off strength; (**c**) view of the discs glued to the coating; (**d**) cohesive and adhesive failure.

3. Results

Due to the different texturing methods of the surface of the concrete substrate, the analysis was carried out separately for each method. The results are the mean values of the pull-off strength and the concrete substrate surface detached thickness, which were obtained for each surface. From all 51 tests, one adhesive failure was observed (in this case 15% of the coating was detached). For the rest of the samples, the cohesive failure was observed in the concrete substrate surface (Figure 8d).

3.1. Pull-Off Strength

It is visible from Figure 9 that only sample No. 1 from the grooving methods obtained values of f_b higher than 1.5 MPa. On the other hand, for all of the imprinting texturing methods, the values of f_b were higher than 1.5 MPa. It was observed that the values of the pull-off strength for sample No. 5 were higher than those declared by the manufacturer (f_b = 2.00 MPa). Half of the patch grabbed samples obtained pull-off strength values higher than 1.5 MPa. The observed values of pull-off strength for brushing were in a range from 1.65 to 1.91 MPa. The pull-off strength result for No. 13 is 1.74 MPa, and for the reference (grinded) surface of the concrete substrate it was equal to 1.82 MPa (No. X).

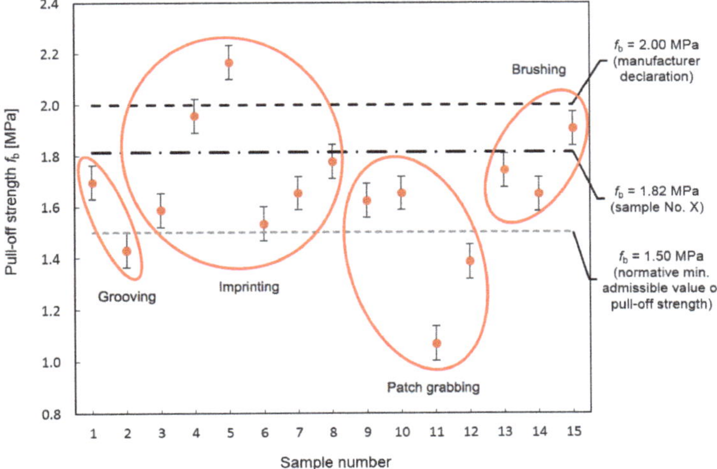

Figure 9. The effect of texturing of the surface of the concrete substrate on the pull-off strength of the epoxy resin coating.

3.2. Detached Thickness

Figure 10 presents the impact of concrete substrate surface detached thickness on the pull-off strength of the epoxy resin coating for grooving (samples from 1 to 2), imprinting (samples from 3 to 8), patch grabbing (samples from 9 to 12) and brushing (samples from 13 to 15).

It is visible from Figure 10a that for the grooved samples the surface of the concrete substrate detached thickness was higher for the painting grid (No. 2) than for the building net (No. 1). Figure 10b shows that in the case of imprinted surfaces, the concrete substrate surface detached thickness increases with the thickness of the cross and the number of crosses that were used on one texturing plank. The greatest result of depth after the pull-off strength test was obtained using the cross with a thickness $b_2 = 6$ mm. The concrete substrate surface detached thickness for the patch grabbed samples Nos. 10–12. increases proportionately to the size of the textured longitudinal stripes (Figure 10c). The concrete substrate surface detached thickness is almost the same as the pull-off strength for samples No. 13 (13.03 mm) and No. X (13.68 mm). Moreover, for the concrete textured by brushing, a smaller concrete substrate surface detached thickness was observed (Figure 10d). The pull-off strength results for the first three methods are the best when the concrete substrate surface detached thickness value is close to 12 mm.

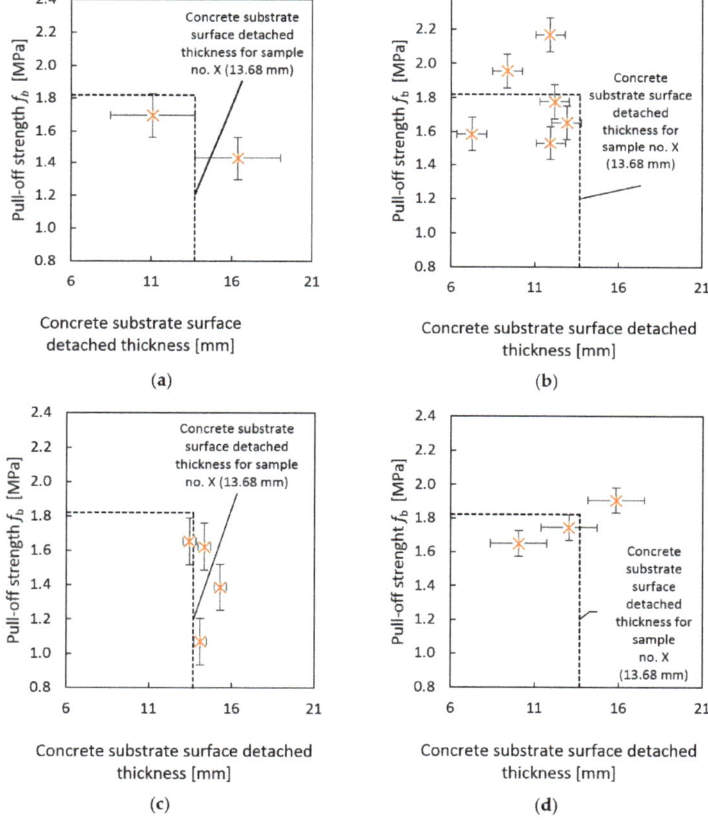

Figure 10. Impact of concrete substrate surface detached thickness on the pull-off strength of the epoxy resin coating for: (**a**) grooving (samples from 1 to 2); (**b**) imprinting (samples from 3 to 8); (**c**) patch grabbing (samples from 9 to 12); (**d**) brushing (samples from 13 to 15).

4. Conclusions

The purpose of this article was to evaluate the effect of texturing of the surface of concrete substrate on the pull-off strength of the epoxy resin coating. The research also aimed to ensure a proper pull-off strength of the epoxy resin coating by preparing the surface of the concrete substrate using different texturing methods. Based on the performed tests, the following conclusions can be drawn:

- The treatment of the surface of the concrete substrate by grinding, cleaning, and applying a primer can be replaced by many different texturing methods that can give similar pull-off strength results.
- Brushing with a painting brush achieved the most similar results to those obtained for the sample after grinding (No. X). It shows that a simple brushing method should be used instead of grinding.
- The best texturing method turned out to be imprinting (No. 5. with crosses type "+"). The pull-off strength for this sample was equal to 2.17 MPa. This is more than that declared by the manufacturer when the surface of a concrete substrate was prepared according to their recommendations (2.00 MPa). Imprinting can be easily used for texturing a concrete substrate surface, as it creates a higher pull-off strength of the epoxy resin coating.
- The concrete substrate surface detached thickness depends on the depth and area of the textured surface. The surface of concrete substrate should be prepared using a method in which the concrete substrate surface detached thickness value is close to 12 mm. This enables the best possible pull-off strength results to be obtained. Such information can be used during the design and construction stage.

The study shows an alternative way to treat the concrete substrate. In further studies, the best concrete substrate texturing methods should be used with modified epoxy resin coating in order to increase the pull-off strength. Some actions should focus on analyzing the concrete substrate surface detached thickness and the impact of texturing depth on the pull-off strength of coatings.

This study also evidenced the technical difficulties for measuring the real adhesion between concrete substrate and epoxy resin. It is proper to note that the pull-off strength test results cannot be used to differentiate the adhesion properties of the epoxy coating to the differently prepared concrete substrates. This is due to the fact that all of the pull-off strength test results resulted in a cohesive failure. Thus, for future studies the development of an adequate method to measure real adhesion between coatings and substrates is required.

Author Contributions: Conceptualization, K.K. and Ł.S.; methodology, Ł.S.; validation, K.K.; formal analysis, Ł.S.; investigation, K.K.; resources, K.K.; data curation, K.K.; writing—original draft preparation, K.K.; writing—review and editing, K.K. and Ł.S.; visualization, K.K.; supervision, Ł.S.

Funding: This research received no external funding.

Acknowledgments: The authors would like to acknowledge the contribution of the COST Action CA15202.

Conflicts of Interest: The authors declare no conflict of interest.

References

1. Gruszczyński, M. The use of thin cement-polymer layers to repair and strengthening concrete floors. *Materiały Budowlane* **2018**, *9*, 30–33. (In Polish) [CrossRef]
2. Sadowski, Ł. *Adhesion in Layered Cement Composites*, 1st ed.; Springer International Publishing: Cham, Switzerland, 2018.
3. Sadowski, Ł. Multi-scale evaluation of the interphase zone between the overlay and concrete substrate: Methods and descriptors. *Appl. Sci.* **2017**, *7*, 893. [CrossRef]
4. Basheer, L.; Kropp, J.; Cleland, D.J. Assessment of the durability of concrete from its permeation properties: A review. *Constr. Build. Mater.* **2001**, *15*, 93–103. [CrossRef]
5. Brown, P.W.; Doerr, A. Chemical changes in concrete due to the ingress of aggressive species. *Cem. Concr. Res.* **2000**, *30*, 411–418. [CrossRef]

6. Safiuddin, M. Concrete damage in field conditions and protective sealer and coating systems. *Coatings* **2017**, *7*, 90. [CrossRef]

7. Rodrigues, M.P.M.C.; Costa, M.R.N.; Mendes, A.M.; Marques, M.E. Effectiveness of surface coatings to protect reinforced concrete in marine environments. *Mater. Struct.* **2000**, *33*, 618–626. [CrossRef]

8. Wang, S.; Li, Q.; Zhang, W.; Zhou, H. Crack resistance test of epoxy resin under thermal shock. *Polym. Test.* **2002**, *21*, 195–199. [CrossRef]

9. Garcia, J.; De Brito, J. Inspection and diagnosis of epoxy resin industrial floor coatings. *J. Mater. Civ. Eng.* **2008**, *20*, 128–136. [CrossRef]

10. Figueira, R.B. Hybrid sol-gel coatings: Erosion-corrosion protection. In *Production, Properties, and Applications of High Temperature Coatings*; Pakseresht, A.H., Ed.; IGI Global: Hershey, PA, USA, 2018; pp. 334–380.

11. Figueira, R.; Callone, E.; Silva, C.; Pereira, E.; Dirè, S. Hybrid coatings enriched with tetraethoxysilane for corrosion mitigation of hot-dip galvanized steel in chloride contaminated simulated concrete pore solutions. *Materials* **2017**, *10*, 306. [CrossRef] [PubMed]

12. Mynarčík, P. Technology and trends of concrete industrial floors. *Procedia Eng.* **2013**, *65*, 107–112. [CrossRef]

13. Sánchez, M.; Faria, P.; Ferrara, L.; Horszczaruk, E.; Jonkers, H.M.; Kwiecień, A.; Mosa, J.; Peled, A.; Pereira, A.S.; Snoeck, D.; et al. External treatments for the preventive repair of existing constructions: A review. *Constr. Build. Mater.* **2018**, *193*, 435–452. [CrossRef]

14. Sadowski, Ł.; Czarnecki, S.; Hoła, J. Evaluation of the height 3D roughness parameters of concrete substrate and the adhesion to epoxy resin. *Int. J. Adhes. Adhes.* **2016**, *67*, 3–13. [CrossRef]

15. Li, Y.; Liu, X.; Li, J. Experimental study of retrofitted cracked concrete with FRP and nanomodified epoxy resin. *J. Mater. Civ. Eng.* **2016**, *29*, 04016275. [CrossRef]

16. Rousakis, T.C.; Kouravelou, K.B.; Karachalios, T.K. Effects of carbon nanotube enrichment of epoxy resins on hybrid FRP–FR confinement of concrete. *Composites Part B* **2014**, *57*, 210–218. [CrossRef]

17. Chowaniec, A.; Ostrowski, K. Epoxy resin coatings modified with waste glass powder for sustainable construction. *Czasopismo Techniczne* **2018**, *8*, 99–109. [CrossRef]

18. Do, J.; Soh, Y. Performance of polymer-modified self-leveling mortars with high polymer–cement ratio for floor finishing. *Cem. Concr. Res.* **2003**, *33*, 1497–1505. [CrossRef]

19. Ahn, N. Effects of diacrylate monomers on the bond strength of polymer concrete to wet substrates. *J. Appl. Polym. Sci.* **2003**, *90*, 991–1000. [CrossRef]

20. Czarnecki, L.; Taha, M.R.; Wang, R. Are polymers still driving forces in concrete technology? In Proceedings of the International Congress on Polymers in Concrete (ICPIC 2018), Washington, DC, USA, 29 April–1 May 2018; Taha, M.M.R., Ed.; Springer: Cham, Switzerland, 2018; pp. 219–225.

21. Bissonnette, B.; Courard, L.; Garbacz, A. *Concrete Surface Engineering*, 1st ed.; CRC Press: Boca Raton, FL, USA, 2015.

22. He, Y.J.; Mote, J.; Lange, D.A. Characterization of microstructure evolution of cement paste by micro computed tomography. *J. Cent. South Univ.* **2013**, *20*, 1115–1121. [CrossRef]

23. Mirmoghtadaei, R.; Mohammadi, M.; Samani, N.A.; Mousavi, S. The impact of surface preparation on the bond strength of repaired concrete by metakaolin containing concrete. *Constr. Build. Mater.* **2015**, *80*, 76–83. [CrossRef]

24. *EN 1542 Products and Systems for the Protection and Repair of Concrete Structures–Test Methods–Measurement of Bond Strength by Pull-Off*; British Standard Institution: London, UK, 2006.

25. *ASTM D4541-95e1 Standard Test Method for Pull-off Strength of Coatings using Portable Adhesion Testers*; ASTM International: West Conshohocken, PA, USA, 2002.

26. Czarnecki, S. Ultrasonic Evaluation of the pull-off adhesion between added repair layer and a concrete substrate. *IOP Conf. Ser. Mater. Sci. Eng.* **2017**, *245*, 032037. [CrossRef]

 *coatings*

Review

Electrochemical Strategies for Titanium Implant Polymeric Coatings: The Why and How

Stefania Cometa [1], Maria Addolorata Bonifacio [1,2], Monica Mattioli-Belmonte [3], Luigia Sabbatini [2] and Elvira De Giglio [2,*]

[1] Jaber Innovation s.r.l., 00144 Rome, Italy; stefania.cometa@jaber.it (S.C.); maria.bonifacio@uniba.it (M.A.B.)
[2] Department of Chemistry, University of Bari "Aldo Moro", 70126 Bari, Italy; luigia.sabbatini@uniba.it
[3] Department of Clinical and Molecular Sciences, Università Politecnica delle Marche, 60020 Ancona, Italy; m.mattioli@univpm.it
*   Correspondence: elvira.degiglio@uniba.it; Tel.: +39-80-544-2021

Received: 1 April 2019; Accepted: 18 April 2019; Published: 20 April 2019

Abstract: Among the several strategies aimed at polymeric coatings deposition on titanium (Ti) and its alloys, metals commonly used in orthopaedic and orthodontic prosthesis, electrochemical approaches have gained growing interest, thanks to their high versatility. In this review, we will present two main electrochemical procedures to obtain stable, low cost and reliable polymeric coatings: electrochemical polymerization and electrophoretic deposition. Distinction should be made between bioinert films—having mainly the purpose of hindering corrosive processes of the underlying metal—and bioactive films—capable of improving biological compatibility, avoiding inflammation or implant-associated infection processes, and so forth. However, very often, these two objectives have been pursued and achieved contemporaneously. Indeed, the ideal coating is a system in which anti-corrosion, anti-infection and osseointegration can be obtained simultaneously. The ultimate goal of all these coatings is the better control of properties and processes occurring at the titanium interface, with a special emphasis on the cell-coating interactions. Finally, advantages and drawbacks of these electrochemical strategies have been highlighted in the concluding remarks.

Keywords: electrochemistry; polymer coatings; titanium implants; corrosion protection; biocompatibility

---

1. Introduction

Electrochemical deposition of polymers (ECD) is a relatively new technique for metal modification, even though since ancient times metals have been coated with non-polymeric films by electrochemical processes (e.g., metal plating, anodization and many others).

On the other hand, the synthesis of organic compounds such as polymers has been traditionally accomplished via chemical routes. Alternatively, over the last century, the use of electrochemical methods for polymer synthesis has been investigated at both the laboratory and industrial scale.

ECD has gathered a considerable consensus, since it combines the advantages of an easily controlled and automated technique with the inherent possibility of coating different conducting or semiconducting substrates with polymers having disparate properties [1]. In general, the polymer ECD can be categorized into two separate methods, schematized in Figure 1:

- Electropolymerization or electrosynthesis: polymer is grown directly on the metal electrode surface, starting with an electrolyte solution containing the relevant monomer. The process can be further divided into potentiodynamic, galvanostatic and potentiostatic electropolymerization.

- Electrophoretic deposition: polymer exists in the form of fine powder or solubilized in the electrolyte solution and it is attracted to the metal electrode due to its intrinsic electric charge.

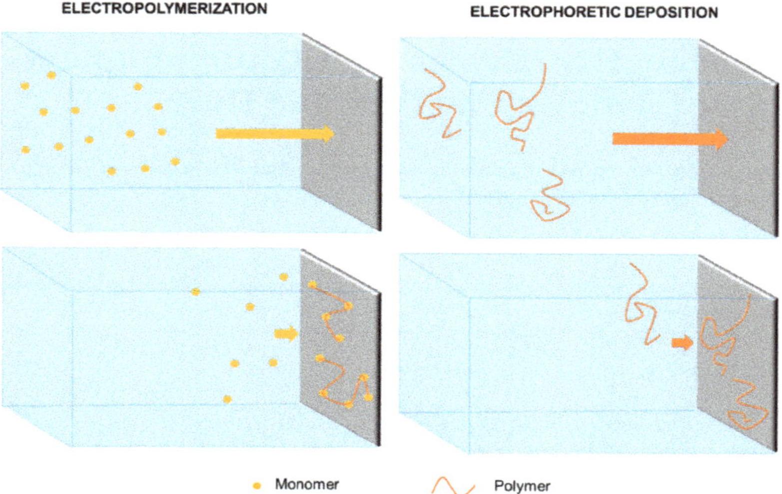

Figure 1. Schematic difference between electropolymerization and electrophoretic deposition of polymers to the metal electrode.

The electropolymerization configuration is based on a three-electrode cell (i.e., working, counter and reference electrodes) containing a solution of the monomer (s) and the electrolyte (dopant) in an appropriate solvent. The monomers can be anodically oxidized or cathodically reduced, forming radical anions or cations that react with other monomers, thus forming on the surface of the involved electrode an insoluble polymer layer, which can be conductive, semi conductive or insulating. For insulating polymers, the thicknesses obtainable by electropolymerization are limited to no more than few hundred nanometres, due to the current drop occurring during the film growth. Electropolymerization allows not only to control the film thickness but also to perform in situ characterization of the polymer during its growth with the use of electrochemical and/or spectroscopic methods. Thanks to these properties, the ECD has resulted particularly fascinating in a wide range of application fields, such as corrosion protection in automotive [2,3], electroanalysis [4], electrocatalysis [5], solar cells [6], electronic and microelectronic devices [7], battery technologies, light emitting diodes, electrochromic displays [8] and so forth.

Surely, the greatest impulse to the electropolymerization was given by the development of a sub-class of the chemical sensors, that is, electrochemical sensors and biosensors. The main conducting polymers employed in biosensing are the polyheterocycles, such as polypyrrole (PPy), polythiophene (PT), polyaniline (PANI) and poly(3,4-ethylenedioxythiophene) (PEDOT), developed in the 1980s.

Noteworthy different examples of PPy-based biosensors have been reported in the literature, thanks to its suitability to immobilize enzymes [9]. The interference-free capability of this electrosynthesised polymer was guaranteed by the overoxidation process, which produced a permselective, antifouling membrane able to reject the other typical serum components. PPy films have been electrosynthesized also on mesoporous titanium oxide and resulted able to successfully immobilize glucose oxidase [10].

As far as the electrophoretic deposition process is concerned, it consists of two steps: electrophoresis, that is, the migration of the macromolecules suspended in a solution toward the electrode, attracted by an electrical field and then the formation of a deposit on the electrode surface [11]. Based on the charge of the electrode and of the macromolecules, two electrophoretic deposition processes can be distinguished: cathodic electrophoretic deposition, when the molecules are positively charged and thus attracted to the cathode; anodic electrophoretic deposition, for negatively charged molecules deposited on the anodic surface.

Both these two main ECD processes allow the irreversible, durable and uniform change of the surface properties of metal substrates. Therefore, ECD finds its natural application in various fields of medicine or biomedical engineering, improving wear resistance or enhancing the affinity of metals with living cells and tissues.

Beyond the classical applications of electrochemical techniques in industrial research areas, the past two decades have seen significant achievements of electrochemistry in biomedicine. Indeed, the development of active materials (e.g., medical devices, modified surfaces), implants, sensors and advanced drug delivery systems has benefited from the knowledge of electrochemical processes naturally occurring in living systems [12]. Moreover, in the last 10 years, electrochemical techniques have been further exploited to create complex patterns of macromolecules, accurately guiding protein deposition on conductive or non-conductive substrates [13]. For instance, the atomic-force-controlled capillary electrophoretic printing (ACCEP) has been developed to control protein positioning with high resolution in time and space, envisioning new opportunities for biomedical research [14]. Furthermore, electrochemistry is also giving a valuable contribution to pharmaceutical applications, especially in the development of remote-control drug delivery systems. In this regard, a noteworthy example was given by Huang et al., who fabricated a flexible antiepileptic-delivery system on PET via electrophoretic deposition. Drug elution was triggered by an external magnetic field, leading to tuneable release kinetics [15]. A similar concept has been exploited to enhance the biocompatibility of metallic cardiovascular devices, mainly made of stainless-steel. Innovative drug-eluting stents (DES) have been developed to in situ release anti-restenotic agents, halving in-stent restenosis phenomena with respect to bare metal devices [16].

Investigations of electrochemical methods is also guiding to very attractive findings in neural repair (e.g., in case of peripheral nerve or spinal cord injuries, glial scar treatment or cochlear functionality restoration). Gomez at al. electrosynthesized PPy layers on gold, in presence of nerve growth factor, observing improved neural cell growth [17]. Furthermore, with a similar approach, Quigley et al. managed to guide Schwann cell migration and axonal growth direction electrodepositing PLA-PLGA aligned fibres on gold substrates [18]. These findings display the potential of electrochemically prepared coatings to modify implant surfaces, aiming at precise cell guidance, with applications in regenerative medicine, especially for electrically-active tissues (i.e., muscular and nervous tissues).

To the best of our knowledge, the longest tradition of implant surface modification with electrochemical techniques is related to orthopaedic and orthodontic devices, with a special focus on titanium and its alloys prosthetic elements (artificial hip joints, artificial knee joints, bone plates, screws for fracture fixation, crowns, bridges, overdentures etc.). Hydroxyapatite layers were already electrodeposited on titanium substrates in 1986 [19]. Since then, a plethora of surface functionalization strategies have been proposed, to endow titanium implants with corrosion resistance, biocompatibility, osseointegration and antimicrobial features. The main phenomena that need to be addressed/avoided when an orthopaedic or orthodontic prosthesis is modified with an ECD coating, are schematized in Figure 2.

A pivotal role of coatings endowed with bioactive, anticorrosion or antimicrobial properties is related to the enhancement of an implant's integration with the surrounding tissues. In this respect, several different in vitro approaches were developed to assess the biocompatibility of electropolymerized or electrodeposited titanium coatings, mainly based on osteoblastic cell lines (i.e., MG63 and Saos-2) or primary cells.

This review article deals with the electrochemical strategies to coat titanium implants with polymeric films, both bioinert and bioactive films. Whereas the former class is mainly focused on titanium-based implants protection against corrosion, the latter involves coatings intended to produce an enhanced biological response, in terms of infection prevention, cell adhesion, new bone matrix deposition and so forth.

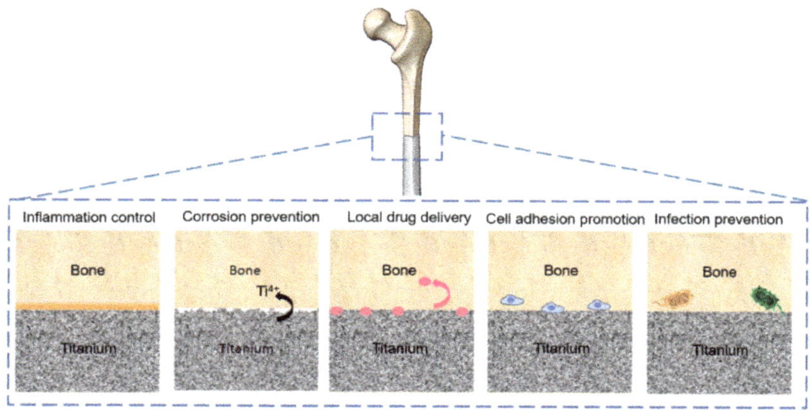

Figure 2. Schematic representation of phenomena occurring at titanium implant-bone interfaces.

In Figure 3, the evolution over time of the number of publications on titanium ECD coatings is reported, suggesting a growing interest toward this research field. On the right of Figure 3, a pie chart shows the distribution of different types of bioactive coatings into three main classes.

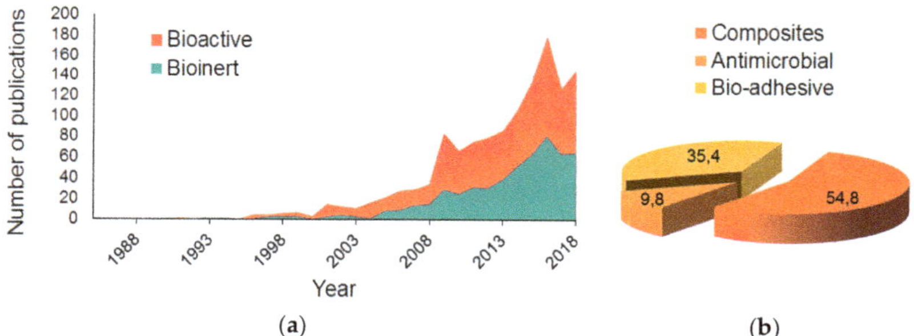

Figure 3. (**a**) Timeline of electrochemical deposition (ECD)-based coatings on titanium distinguished in bioinert and bioactive polymer films. (**b**) Pie chart relevant to the different bioactive coatings. Data source: 2019 Scopus®.

The State-of-art in this research field is summarized with the aim to gain insight into the future perspectives of ECD in biomedical fields.

2. Bio-Inert Systems: Anticorrosion and Barrier Films

Corrosion of metallic biomaterials is the gradual degradation of the metal surface by an electrochemical attack, which occurs when the metallic implant is placed in the hostile, highly oxygenated, saline, electrolytic environment of the human body. Blood and other constituents of body fluids are rich of different corrosive substances, including water, amino acids or proteins, plasma, mucin in the case of saliva [20], various anions (such as chloride, phosphate and bicarbonate ions), cations (like sodium, potassium, calcium, magnesium, etc.), organic substances of low-molecular-weight species as well as relatively high molecular-weight polymeric components and dissolved oxygen [21,22]. Changes in the pH values also have a deep impact on metal implant corrosion. Human body pH value is normally near 7.0; however, this value could drastically decrease due to diseases, infections and other factors and after surgery. Indeed, the pH value near the implant could lower up to 5.5 or 5.0. Even if metal implants are routinely pre-passivated prior to final packaging using electrochemical methods [23], a clinical evidence for the release of metal ions from the implants has been established and this leaching has been ascribed to corrosion processes. Obviously, corrosion attack on metal components of surgical implants is one of the main factor responsible of their short lifetime. This issue is even more severe in the case of prostheses with mobile components. For example, life expectancy of a hip implant is no more than 10 years, mainly due to corrosion phenomena accelerated by the mechanical wear [24].

Commercially pure titanium (CpTi) and its alloys are widely used in orthopaedic and orthodontic applications, also due to their acceptable corrosion resistance [25]. Despite titanium-based implants might not cause any biological adverse reaction, long term stability of these prostheses/implants cannot be totally guaranteed. The lack of surface chemical-physical stability due to corrosive processes could produce many side effects. Indeed, Olmedo and co-workers [26] evidenced the relation between peri-implantitis, a site-specific infectious disease and ion-induced corrosion of the titanium surface.

Resistance to corrosion of titanium implants can be increased by alloying it with aluminium and vanadium or aluminium and niobium (even if also other metals such as molybdenum, zirconium, rhenium, chromium, nickel or manganese have been used). The most frequent use of the Ti-6Al-4V alloy for either orthopaedic or orthodontic implants is linked to an interesting combination of resistance to corrosion, durability, low elasticity module and high osseointegration [27–29]. However, several issues can be related to the side effects of the alloy's components. Detailed studies have shown that they lead to long term ill effects such as peripheral neuropathy, osteomalacia and Alzheimer disease due to the release of aluminium and vanadium ions from the alloy. In addition, vanadium, which is present both in the elemental state and in oxides (V_2O_5) is highly toxic [30].

In recent years, various researchers have studied different strategies to mitigate titanium-based implants corrosion. In general, this could be attempted by surface modification techniques [31]. Some examples of surface modification processes are physical and chemical vapour deposition [32], laser treatments [33–35], thermal oxidation [36,37] and thermal spraying [38], plasma spray [39,40], ion implantation [41], micro-arc oxidation [42], sandblasting [43,44] and electrochemical treatment [45].

Among these surface modification techniques, the manufacture of multi-materials obtained by deposition of coatings on metals can be a successful way for the mitigation of the metal corrosion process. Polymers, in particular, represent an optimal building block for the achievement of a barrier film on the metal surface.

Different coating strategies can be adopted, such as plasma assisted deposition [46], solvent evaporation [47], sol-gel dip coating [48], dip coating [49] and so forth. As a matter of fact, the application of polymeric protective coatings via electrochemical processes on titanium represents one of the most powerful strategies to obtain a superior adhesion of the polymer layer to the metal substrate with high resistance to mechanical wear.

Although this approach has been widely used in the development of anti-corrosion coatings on various metals (iron, steel, copper, etc.) in different sectors such as naval, aerospace, automotive and so forth [50–52], in the biomaterials field the development of electrochemical polymeric coatings on metals and, in particular, on titanium based implants, for corrosion protection remains a niche study.

In this respect, one of the first studies was carried out by our research group, which achieved the electrosynthesis of a polypyrrole (PPy) coating in aqueous media on both titanium and TiAlV alloy [53]. It is worth noting that, while PPy films are easily synthesized at inert anodes, the electropolymerization at oxidizable metals can occur only under electrochemical conditions that strongly passivate the electrodes without preventing the electropolymerization growth. Indeed, PPy coating is poorly adherent to the oxidizable metals (Fe, Zn, Al, Ti) because of the competition between two simultaneous anodic processes: PPy formation and metal oxidation. Interestingly, in this work it was demonstrated that a thin layer of the native oxide film did not hinder an efficient PPy polymerization on the titanium surface. This polymer is particularly versatile in that it can be either easily functionalized with biologically active molecules, able to stimulate positive interactions with bone tissue or act as an efficient protective barrier. For example, PPy was polymerized starting from the relevant monomer solution on anodically polarized NiTi electrode, demonstrating that the polymer improves the corrosion performance at the open circuit potential and at potentials where the bare substrate suffers pitting attack [54]. An improvement in the corrosion resistance property was attempted by the electro-co-deposition of PPy and PEG (composite coating) on Ti-6Al-7Nb alloy. The authors in this paper substantiated the crucial role of PEG in the corrosion tests carried out in Hank's balanced salt solution [55]. The biological response of G292 human OBs was tested onto these PPy/PEG films, showing a good cell viability and proliferation. Furthermore, G292 preserved their morphology within the first 24 h of culture. Overall, the study suggested that the proposed film offers a microenvironment with an increased bioactivity, improved corrosion resistance and biocompatibility. Finally, another paper proposed the modification of Ti-based implants surfaces through incorporation of torularhodin, a natural compound with antimicrobial effect, by means of a polypyrrole film. The results showed that PPy–torularhodin composite film, besides showing antibacterial activity and no harmful effect on cell viability, acted also as an anticorrosion coating [56]. These observations, as well as other studies with different cytotypes [57], strengthened the role of PPy as a material for different biomedical applications.

Another class of polymers showing interesting barrier properties are the polyacrylates. In 2005, our research group focused on the study of performances of poly(methyl methacrylate) (PMMA) coatings, as such or modified by an annealing process, as barrier films against corrosion of titanium-based orthopaedic implants [58]. In this work, the electrosynthesis of MMA on titanium substrates was performed for the first time by an electro-reductive process from aqueous solutions. This study evidenced that the presence of PMMA coatings produced a decrease in ion release from Ti alloys. Moreover, the annealing treatment considerably reduced the ion dissolution rate, leading to very efficient protective coatings. As a further improvement of this research, De Giglio et al. carried out the electrosynthesis of poly(acrylic acid) (PAA) films on pure Ti or Ti-6Al-4V sheets [59]. The idea was to obtain a versatile coating for titanium-based orthopaedic implants acting both as an effective anti-corrosion barrier and as a bioactive surface, thanks to the presence of carboxylic groups in PAA that can be functionalized with bioactive molecules. Also in the case of PAA coatings, the annealing procedure resulted in a more compact film able to strongly inhibit the ion release, as demonstrated in simulating tests. Since PAA is a poly-anion, the pH dependency of the polymer barrier properties was also studied [60], indicating that PAA barrier properties were optimal at a pH value near the polymer pKa (4.9); this could be ascribed to a good compromise between the number of charged groups and the polymer swelling.

More recently, Meng et al. [61] synthesized a copolymer Poly(2-Hydroxyethyl methacrylate-*co*-2-(Dimethylamino) ethyl methacrylate-*co*-7-hydroxy-4-methylcoumarin methacrylate) (PHDC))–by free radical polymerization and then self-assembled into colloidal particles and immobilized on the NiTi alloy by a simple one-step electrophoretic deposition. This study demonstrated that the copolymer coatings could significantly decrease the release of nickel ions into the environment. As far as the coating's biological evaluation is concerned, two different cytocompatibility approaches were tested. Indeed, NIH 3T3 fibroblasts were used to assess both the effect of the release of nickel ions through the physical coating barrier and the effect of the direct interaction between the cells and the coating.

Another promising polymer in the biomaterials field is poly(ether ether ketone) (PEEK). The most important properties of PEEK are low density, high mechanical and tribological properties, as well as good chemical and thermal stability and irradiation sterilization resistance. It was also exploited to develop coatings, as such or reinforced with different fibres or particles, by electrodeposition on Ti-13Nb-13Zr [62,63]. Interesting results were obtained in terms of improvement of tribological properties, such as wear resistance, friction coefficient and adhesion of the coatings to the substrate, as well as corrosion resistance. Indeed, scratch tests have been performed on these coatings, revealing no delamination even up to the maximal load of 30 N (Figure 4a). Cracks at the borders and within the scratch track were visible at high magnifications (Figure 4b). Overall, this analysis evidenced an excellent adhesion of PEEK coatings to Ti-13Nb-13Zr.

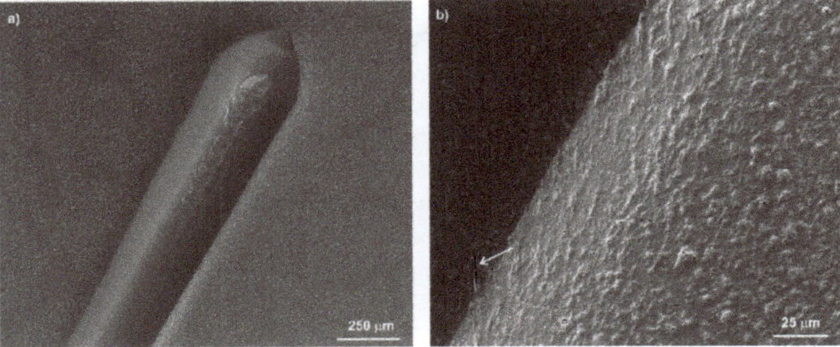

Figure 4. Scanning electron microscope (SEM) images of the poly(ether ether ketone) (PEEK) coating on Ti-13Nb-13Zr after scratch test at a load of 30 N (**a**) and 25 N (**b**). (Reprinted with permission from [63] 2016 Elsevier, copyright n. 4557620529187).

Composite coatings, based on bioglass/chitosan and sol-gel glass/chitosan, were electrophoretically deposited on a near-β Ti-13Nb-13Zr alloy [64]. It was found that both types of coating improve the electrochemical corrosion resistance of the Ti-13Nb-13Zr alloy in Ringer's solution. Furthermore, a good biocompatibility was demonstrated with MG63 osteoblast-like cells.

A recent work discussed the electropolymerization of poly(3,4-ethylenedioxythiophene) (PEDOT) coating on near-β Ti-20Nb-13Zr (TNZ), which was pre-treated with three different surface treatments to ameliorate the substrate in terms of morphology, topography and hydrophobicity as well as to facilitate the formation of a compact PEDOT film [65]. The results from electrochemical corrosion studies clearly evidenced that the PEDOT coatings on the surface-treated TNZ substrates improved barrier protection performances in simulated body fluid (SBF).

A further example of protective polymer film electrodeposited on titanium substrates was reported by Bosh et al. [66], which deposited on Ti an anti-corrosion coating based on Halar®, an ethylene chlorotrifluoroethylene (ECTFE) thermoplastic polymer, followed by post heat treatment. It was concluded that this polymer could be applied by electrophoresis as protective coating to improve the physical and mechanical properties of the metal substrate and to reduce the stress shielding effect.

Finally, a recent paper reviewed the corrosion resistance of the main types of biocompatible metals, declaring that metals will reasonably continue to be used as biomaterials due to their unmatched mechanical excellence [67]. Therefore, electrochemical polymeric coatings can still provide new solutions for improving the stability of titanium prostheses.

3. Bioactive Coatings: From Drug-Eluting Systems to Antimicrobial Surfaces

The previous section dealt with the strategies to protect titanium implants from unwelcome phenomena (e.g., corrosion) due to the interaction between the biomaterial and the surrounding tissues. Besides, several researchers are also focusing on a complementary issue, that is, implant integration through the enhancement of cell adhesion and new matrix deposition.

The interface between a biomaterial and the host microenvironment is the main player in the process that determines implant success or failure. Indeed, several strategies are focusing on titanium surface modification in order to improve cell colonization [68].

An effective method consists in providing titanium with topographical features that stimulate cell adhesion. As an example, Wang and co-workers combined sandblasting, anodic oxidation and acid etching to fabricate micro- and nanoscale structures on titanium implants, studying the effect on cell behaviour [69]. Several other papers [70–73] proposed different techniques (ion implantation, plasma spray, lithography, electric field-aided casting) to engineer metallic surfaces, creating hierarchical patterns to guide cell adhesion. Jeon et al. discussed the impact of pattern shapes (e.g., disordered squares, grooves, pillars) on different cell types, providing new perspectives on implant fabrication to finely tune cells' activities [74].

Moreover, beyond the choice of a specific surface pattern, the achievement of a bioactive implant could take advantage from the modification of its surface's physico-chemical features (e.g., wettability, protein adsorption features, degradation rate). Indeed, bare titanium and its alloys, as well as other metallic biomaterials, do not display excellent "biofunctions" intended as the abilities to attract cells and being spontaneously integrated in the surrounding tissues [75]. In this context, the development of bioactive polymeric coatings represents one of the most effective approaches available at reasonable costs. The electrochemical process allows the direct growth of polymeric coatings on titanium, leading to tightly adherent films with adjustable thickness [76]. In this respect, our research group prepared poly(2-hydroxyethyl methacrylate) (PHEMA) coatings on different metallic substrates, using cyclic voltammetry. By simply varying the number of cycles or adding ethylene glycol dimethacrylate (EGDMA) as crosslinker, the authors highlighted the opportunity to tune the polymer thickness, porosity and swelling characteristics, affecting cell adhesion and morphology. The biocompatibility of PHEMA was assessed with fibroblasts, evidencing that hydrogel surfaces had a strong effect on cell adhesion, thus proving useful for both biomedical and diagnostic devices. Indeed, the crosslinked PHEMA-EGDMA coating displayed a smooth and continuous surface, which promoted fibroblasts elongation, while the uncrosslinked PHEMA coating, with its more porous structure, improved cell adhesion and spreading. More recently, Bhattarai et al. electrosynthesized a poly(aniline) coating via cyclic voltammetry, using a titanium-anodized substrate with nanotubular geometry [77]. Their work demonstrated the role of the polymeric coating in enhancing pre-osteoblasts attachment, spreading and osteogenic differentiation. Moreover, Popescu et al. afforded the deposition of a PPy coating through the bioinspired self-polymerization of an adhesive poly(dopamine) layer, which was able to improve the stability and duration of the PPy coating [78].

Beyond the electrosynthesis of polymers starting from their relevant monomers, electrophoretic deposition is an effective alternative, widely used to prepare films based on nature-derived polymers. In this respect, Kamata et al. developed a collagen-electrodeposited coating on titanium, discussing the superior homogeneity of electrochemically assisted deposition as compared to a conventional dipping technique [79]. Moreover, Zhuang et al. recently managed to align chitosan nanofibers on titanium using a magnetically assisted chemical electrodeposition that involved the incorporation of iron oxide nanoparticles (IOPs) and the application of an external magnetic field during ECD. This aligned collagen coating positively guided bone marrow mesenchymal stem cells (BMSCs) growth and elongation, favouring cell differentiation toward an osteogenic phenotype (e.g., expression of *m*RNA for ALP, COLL I and OCN) [80].

In addition to the already discussed benefits, coating of metallic implants with a polymeric, bioactive film involves the advantage to use it as a versatile platform to load and deliver bioactive molecules (e.g., drugs, growth factors or proteins). Through electrochemical techniques, the bioactive molecules could be entrapped during the coating formation and their release kinetics could be controlled varying the coating's features, as well as the electrochemical parameters. Moreover, since biomolecules are often sensitive to harsh temperatures, extreme pH conditions and UV exposure, the electrochemical route could preserve them from denaturation or chemical rearrangement.

Following these principles, in 2008 our research group electrosynthesized three acrylate-based hydrogels in presence of a model drug (caffeine) and a model protein (bovine serum albumin), pointing out that their release could be tuned by controlling the nature and/or the amount of the coating's crosslinker [81].

Several classes of bioactive molecules (e.g., anti-inflammatory and anticancer drugs, antimicrobials, growth factors) have been loaded on titanium and its alloys, exploiting polymeric coatings to modulate their release kinetics. As an example, indomethacin, an anti-inflammatory drug, was loaded on titanium, modified with an array of titania nanotubes (TNT) and coated with a thin, biodegradable layer of chitosan and poly(lactide-co-glycolide) [82]. The presence of the polymeric coating enabled to extend the drug release up to one month and to adjust it, playing with the film thickness. Furthermore, the model drug ibuprofen was also loaded in an electrochemically deposited chitosan coating on titanium by means of an inorganic carrier (i.e., mesoporous silica nanoparticles) [83] (Figure 5).

This system demonstrated different drug release profiles in response to pH and electrical stimuli, providing new opportunities to trigger the desired drug release with externally controlled signals. In addition, to address the complex issue of poor vascularization around the implant, other authors managed to develop a VEGF-loaded coating on titanium implants, able to promote both mineralization and in situ angiogenesis, two processes strictly interrelated during osseointegration. MG63 human osteoblast-like cell behaviour was assessed up to 21 days, demonstrating their adhesion, viability and maintenance of osteoblastic phenotype [84,85].

Even if the controlled release of a bioactive molecule is the goal to achieve in several applications, a wide range of surface functionalization techniques is also devoted to the development of stable, bioactive surfaces via grafting of chemical groups (e.g., moieties to be used as organic bridges to improve titanium reactivity, adhesive molecules, anti-thrombogenic drugs, etc.). In this respect, De Giglio's research group focused on the electrochemical coating of titanium with a tightly adherent PPy film, modified with L-cysteine as terminal residue [53,86].

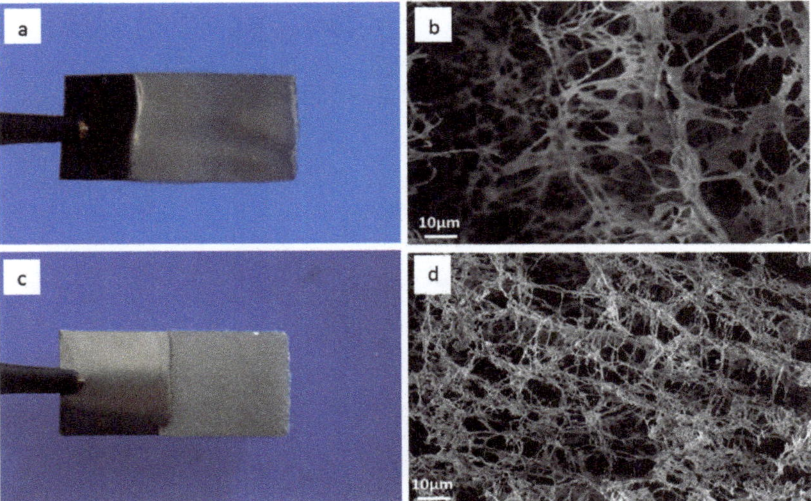

Figure 5. Optical picture (**a**) and SEM image (**b**) of electrodeposited chitosan without ibuprofen-mesoporous silica nanoparticles. Optical picture (**c**) and SEM image (**d**) of the chitosan coating embedding ibuprofen-mesoporous silica nanoparticles. (Reprinted with permission from [83] 2014 Elsevier, copyright n. 4557620011536).

Furthermore, this single amino acid was exploited to covalently bind a peptide sequence containing the RGD moiety, to enhance osteoblasts attachment on the implant surface [87]. Indeed, the authors observed an adhesion improvement of newborn rat calvaria cells equal to 230% on the modified coating, correlated to RGD surface density. Other authors grafted the RGD sequence on PEG coatings, exploiting different techniques (plasma polymerization, electrodeposition, silanization). Their aim was to achieve an antifouling system able control osteoblasts adhesion while inhibiting random protein adsorption and biofilm formation. Morphological observation showed that RGD improved cells adhesion, whilst no differences were detected for cells spreading and morphology. Authors concluded that the impaired cell adhesion due to the antifouling feature of the PEG coatings was improved by the immobilization of the RGD cell adhesive motive, while maintaining the effect on the bacterial adhesion [88].

Another successful strategy consists in modifying the relevant monomers for electropolymerization, in order to confer further reactivity to the coating. In this regard, a carboxylic acid-substituted PPy (PPy-3-acetic acid) was electrosynthesized on titanium, allowing the subsequent grafting of adhesive sequences [89]. Cell adhesion, which is essential in cell growth, migration and differentiation, as well as mRNA levels of Alkaline Phosphates (ALP), Collagen Type I (COLL I) and osteocalcin (OCN) were studied up to 4 days of cell culture. The viability of osteoblasts (OBs), obtained from mouse bone marrow, was not affected by the developed coating. Moreover, OBs adhered and proliferated on the electrosynthesized PPy-3-acetic film, preserving their osteoblastic phenotype.

Covalent binding techniques are also being investigated to graft anti-thrombogenic molecules on metallic implants, hindering platelets activation. As an example, Li et al. proposed a fibronectin-heparin coating on titanium, able to accelerate the endothelialization process, improving blood compatibility of metallic implants such as heart valves, stents or ventricular pumps [90]. In addition, Wagner and co-workers explored the opportunity to lower the thromboembolic risk associated with titanium implants used as ventricular assist devices. These authors modified titanium surfaces with poly(2-methacryloyloxyethyl phosphorylcholine) (MPC), a phospholipidic polymer. They demonstrated superior in vitro blood compatibility, opening new perspectives for milder intravenous anticoagulant therapies [91].

All the examples cited so far demonstrate that the combination of polymeric coatings with metallic biomaterials helps overcoming the drawbacks related to bare metals. As far as the bioactivity of a metallic implant is concerned, one of the most successful strategies is based on the development of composite biomaterials, able to mimic natural tissues. In this respect, composite biomaterials are particularly useful for orthopaedic applications, given the intrinsic dual nature of bone, with its organic (mainly collagen) and inorganic (hydroxyapatite) phases tightly intertwined. Several attempts to recreate native bone microenvironment on titanium implants have been reported, consisting in immobilizing the main components of the surrounding tissue on the biomaterial surface [92–94]. As an example, collagen was embedded during PPy electropolymerization, adjusting the applied potential to allow the collagen fibres, positively charged, to be attracted by a titanium electrode. Then, hydroxyapatite was sprayed on the PPy coating leading to a bone-mimicking composite surface [95]. More recently, other authors observed that the use of bioinspired materials (e.g., nacre, bioactive glasses, silk fibroin, etc.) could stimulate implant integration better than the precise reproduction of native bone. Therefore, several research groups studied hydroxyapatite substitutes to trigger an improved osteointegration of metallic implants [64]. Similarly, collagen was replaced by other low-cost nature-derived polymers, such as chitosan. Recently, Zhang et al. reported the successful deposition of calcium phosphate/chitosan/gentamicin films on titanium alloys, enhancing bone regrowth at levels significantly higher than the normal bone growth rates [96]. Moreover, Boccaccini's research group developed a chitosan/bioactive glass composite on titanium alloys, previously treated by grit blasting in order to tune the surface topography, roughness and wettability. The authors discussed the opportunity to change the applied voltage during electrophoretic deposition to modify the coatings' morphology [97]. With the same perspective, Metoki et al. chemisorbed on Ti-6Al-4V several self-assembled monolayers (SAM) with different end-group charge, length of the chains and anchoring groups, observing their impact on the electrodeposition of calcium phosphate (CaP). The morphology of the coating was mainly affected by the end-group type of the SAM, affecting cell colonization, spreading and biomineralization around the implant [98].

The reviewed strategies to promote biomaterials' osseointegration are often combined with tools to prevent implant infections. Indeed, a delicate, dynamic equilibrium between the risk of bacterial colonization and host's cells adhesion is immediately established after implantation, a process which is called "the race for the surface" [99]. Several works have been addressed to investigate if polymeric coating endowed with antimicrobial properties can be tethered to biomaterial surfaces without losing functionality as well as maintain compatibility with the surrounding environment. Such strategies allow the administration of the antimicrobial agents with the benefit of providing direct interaction with the site of infection. Different strategies could be pursued, such as antibiotics loading into swellable polymers, release during polymer degradation and/or the use of antibacterial molecules directly as coating components. An example of antimicrobial compounds loaded into a swellable polymer is reported in reference [100].

In this respect, electropolymerized coatings based on PHEMA swelling were designed to deliver antimicrobial drugs. Cell adhesion and viability were evaluated up to 120 h of culture, considering the burst release of the ciprofloxacin from the proposed system. Both cytoskeletal organization and cell morphology were unaffected by the presence or release of ciprofloxacin as evidenced by immunofluorescence and SEM observations (Figure 6). Moreover, Cometa et al. electrosynthesized a poly(ethylene glycol diacrylate) (PEGDA) coating on titanium, entrapping vancomycin or ceftriaxone onto its surface. The coatings inhibited methicillin-resistant *Staphylococcus aureus* in vitro, providing a useful tool for in situ prevention of dental or orthopaedic infections [101].

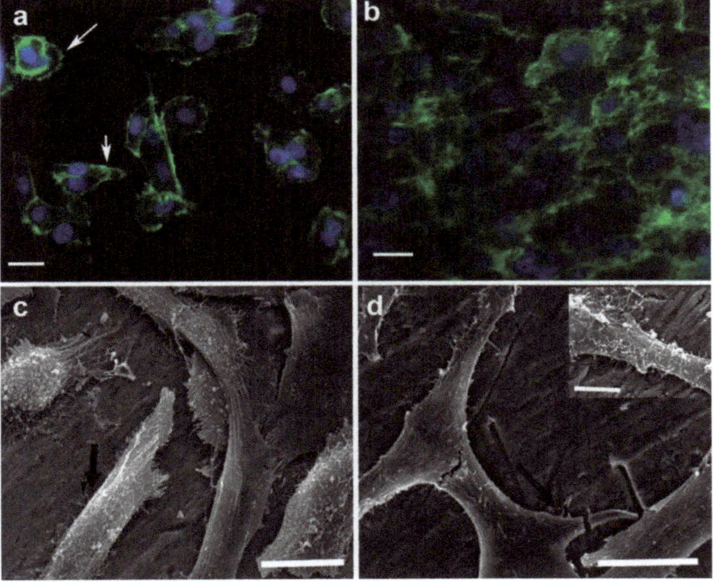

Figure 6. Fluorescence detection of F-actin stress fibres (**a,b**) and SEM micrographs (**c,d**) of MG63 cells cultured on ciprofloxacin-loaded PHEMA coatings on titanium at 24 h (**a,c**) and 120 h (**b,d**) of culture. Arrows in (**a**) indicate local contacts, in (**c,d**) filopodia and a star-shaped morphology. Main figure scale bar 20 μm; inset scale bar 5 μm. (Reprinted in part with permission from [100] 2011 Elsevier, copyright n. 4557620729304).

More recently, Raj et al. developed a titanium alloy coating based on TiO_2-SiO_2 mixtures, chitosan-lysine biopolymers and electrodeposited gentamicin sulphate (Figure 7). The antimicrobial experiments against *S. aureus* and *E. coli* demonstrated that the coating could be useful to eradicate bone infections caused by Gram-negative and Gram-positive bacteria, while repairing bone loss subsequent to osteomyelitis [102].

Beyond conventional antibiotic drugs, antimicrobial peptides (AMP) represent an intriguing alternative to address the concerning issue of drug resistance. In this regard, Hoyos-Nogués et al. prepared a trifunctional coating based on a PEG anti-fouling polymer, embedding both cell adhesive molecules and an AMP. The modified titanium surface hindered protein adsorption and *S. sanguinis* adhesion, while promoting the attachment and spreading of osteoblasts. Data reported are within the 4 h after seeding and are promising for future application in bone replacing approaches [103].

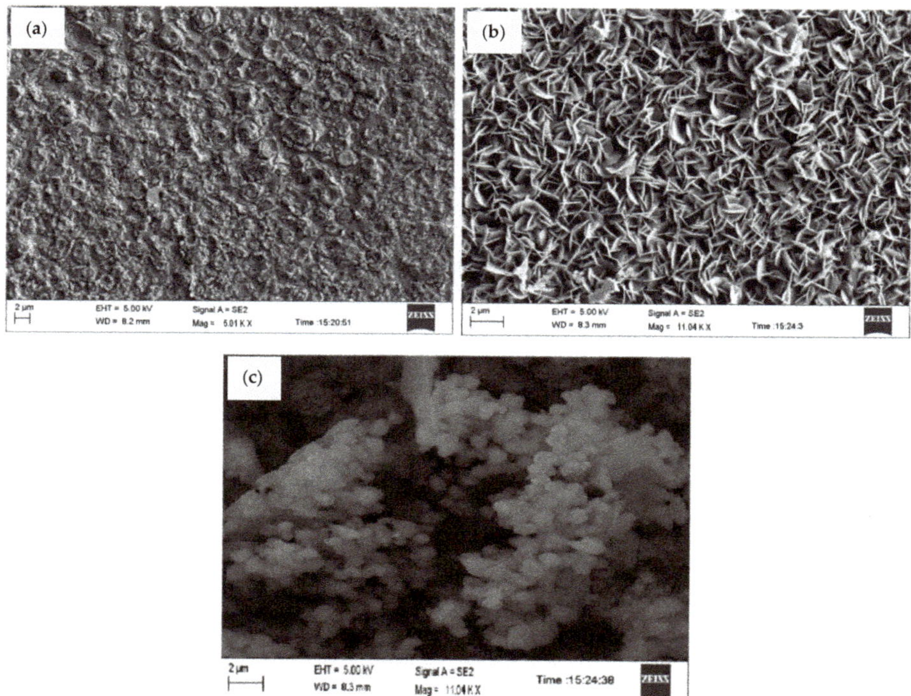

Figure 7. FESEM images of TiO_2–SiO_2 coating (**a**), TiO_2–SiO_2/Chitosan-Lysine coating (**b**) and hydroxyapatite formation on Drug loaded on TiO_2–SiO_2/Chitosan-Lysine coating (**c**). (Reprinted with permission from [102] 2018 Elsevier, copyright n. 4557620827469).

Even antimicrobial metals, in the form of ions or nanoparticles, have been extensively studied as antibiotics substitutes, because of their ability to inhibit microbial proliferation and biofilm formation [104,105]. Several studies pointed out the effectiveness of silver ions and/or nanoparticles, even if cytocompatibility was not always considered [106]. In 2013, De Giglio et al. electrosynthesized a PEGDA-co-acrylic acid coating, embedding silver nanoparticles obtained by a green procedure. The developed system displayed in vitro antibacterial effectiveness on two *S. aureus* clinical isolates. The presence of AgNPs showed no significant toxic effects on osteoblast-like cells in a week of exposure to the examined coatings [107]. More recently, the same authors prepared composite polymeric coatings based on poly(acrylic acid) and chitosan, modified with silver or gallium ions to merge antibacterial effectiveness with osteointegration enhancement [108,109]. Authors evidenced that the viability of MG63 was slightly affected by the gallium-loaded bilayer, whilst no differences were present for cell adhesion, morphology and maintenance of phenotype. Interestingly, changes in coating topography due to the chemical composition represented an instructive pattern for cell arrangement. As far as the silver-loaded bilayer was concerned, cells were affected by the presence of Ag within the polymeric matrix only during the first hours of contact, whilst good cell adhesion, morphology and viability were detected after 7 days of culture. This suggests an acceptable compatibility of the silver-modified coatings with osteoblast-like cells. Overall, the biological observations strengthened the effectiveness of these bilayer coatings for a potential application in orthopaedic and/or dental field. An agent-based modelling approach was also adopted to simulate *S. aureus* interactions with silver-loaded titanium, supporting the choice of the least effective amount of the antibacterial agent [110]. Furthermore, Kim and co-workers electropolymerized poly(dopamine) films, achieving the spontaneous reduction of silver, thus leading to an effective antimicrobial coating [111]. The electrophoretic deposition approach was also exploited by Eraković et al. to attract a silver-doped hydroxyapatite, dispersed in lignin, onto a titanium electrode. Applying constant voltage, it was possible to obtain a system with immediate and continuous release of silver ions, enabling the inhibition of *S. aureus* growth. Moreover, the coating did not show toxicity against peripheral blood mononuclear cells [112].

The growing research on metal-based antimicrobial coatings suggests that the latter would be potentially coupled with conventional antibiotics to overtake drug resistance, exploiting the effectiveness of metals while limiting their toxicity.

4. Conclusions and Future Perspectives

ECD is a fascinating technique for the modification of different metal surfaces and, in particular, it has been successfully applied in biomaterials research, thanks to the possibility of using metals currently employed in both orthopaedic and orthodontic fields as working electrodes.

Not so many techniques are characterized by the main advantages of the ECD, such as low costs, high purity of products, high deposition rates, homogenous covering distribution, capability of combining different monomers to obtain copolymers with a wide range of composition (in the case of electropolymerization), possibility of employing complex geometries of substrate, reduced waste materials, ease of process control and automation.

Moreover, polymer coatings with customized morphology and assemblies can be simply obtained by controlling ECD operating conditions (i.e., initial pH, electrolytic solution composition, current density, deposition time and temperature, typology of electrochemical cell or experimental design etc.).

Finally, ECD can be performed at room temperature from water-based electrolytes, avoiding the use of organic solvents. Hence, ECD can be considered a green chemistry approach to be employed without concerns in biomaterials field, where it is mandatory to avoid the use of potentially toxic substances.

However, there are still some issues to be solved, that represent the current research challenges. Mainly, both electropolymerization and electrodeposition usually produce a fairly dense polymer. Hence, the growth of a passivating polymer film prevents a further polymer deposition (in electrodeposition) or monomer migration (in electropolymerization) on the electrode surface, leading only to very thin coatings. Consequently, other techniques are nowadays required to develop highly porous structures or to obtain thick coatings on titanium.

For all these reasons, when ECD strategies can be exploited, then relatively small areas of conductive substrates could be coated with polymers. Therefore, rather than the scale up of these procedures to coat metal surfaces of large size, the future perspectives of ECD will be related with miniaturisation. Certainly, for the biomedical research, as well as for other technological sectors, it is predictable that the major future trend will be the miniaturisation, rather than the use of macro-devices. In this respect, electrochemical deposition of polymers on metals represents an intriguing strategy, due to the possibility to operate with microelectrodes.

Finally, in the near future, ECD applications are expected to produce tangible improvements in several technological fields, including anti–static coatings, transistors, electromagnetic shielding, organic radical batteries and so on. It is likely that, in the future, the choice of polymeric materials to be employed will fall more often on materials obtained from renewable sources rather than from petrochemical origin. Therefore, with all the advantages discussed in this review, ECD techniques will lead the open challenge to afford "totally green" polymeric coatings on metallic substrates.

Funding: This research was funded by University of Bari "Aldo Moro".

Conflicts of Interest: The authors declare no conflict of interest.

References

1. Beck, F. Electrodeposition of polymer coatings. *Electrochim. Acta* **1988**, *33*, 839–850. [CrossRef]
2. Mengoli, G.; Musiani, M.M. An overview of phenol electropolymerization for metal protection. *J. Electrochem. Soc.* **1987**, *134*, 643C–652C. [CrossRef]
3. Mengoli, G.; Bianco, P.; Daolio, S.; Munari, M.T. Protective coatings by anodic coupling polymerization of o-allylphenol. *J. Electrochem. Soc.* **1981**, *128*, 2276–2281. [CrossRef]
4. Kalimuthu, P.; John, S.A. Electropolymerized film of functionalized thiadiazole on glassy carbon electrode for the simultaneous determination of ascorbic acid, dopamine and uric acid. *Bioelectrochemistry* **2009**, *77*, 13–18. [CrossRef]
5. Xiao, L.; Wildgoose, G.G.; Compton, R.G. Exploring the origins of the apparent "electrocatalysis" observed at C60 film-modified electrodes. *Sens. Actuators B* **2009**, *138*, 524–531. [CrossRef]
6. Granqvist, C.G. Transparent conductors as solar energy materials: A panoramic review. *Sol. Energy Mater. Sol. Cells* **2007**, *91*, 1529–1598. [CrossRef]
7. Wang, D.W.; Li, F.; Zhao, J.; Ren, W.; Chen, Z.; Tan, J.; Wu, Z.; Gentle, I.; Lu, G.Q.; Cheng, H.M. Fabrication of graphene/polyaniline composite paper via in situ anodic electropolymerization for high-performance flexible electrode. *ACS Nano* **2009**, *3*, 1745–1752. [CrossRef]
8. Reiter, J.; Krejza, O.; Sedlaříková, M. Electrochromic devices employing methacrylate-based polymer electrolytes. *Sol. Energy Mater. Sol. Cells* **2009**, *93*, 249–255. [CrossRef]
9. Foulds, N.C.; Lowe, C.R. Enzyme entrapment in electrically conducting polymers. Immobilisation of glucose oxidase in polypyrrole and its application in amperometric glucose sensors. *J. Chem. Soc. Faraday Trans. 1* **1986**, *82*, 1259–1264. [CrossRef]
10. Cosnier, S.; Senillou, A.; Grätzel, M.; Comte, P.; Vlachopoulos, N.; Renault, N.J.; Martelet, C. A glucose biosensor based on enzyme entrapment within polypyrrole films electrodeposited on mesoporous titanium dioxide. *J. Electroanal. Chem.* **1999**, *469*, 176–181. [CrossRef]
11. Van der Biest, O.O.; Vandeperre, L.J. Electrophoretic deposition of materials. *Annu. Rev. Mater. Sci.* **1999**, *29*, 327–352. [CrossRef]
12. Lewenstam, A.; Gorton, L. *Electrochemical Processes in Biological Systems*; John Wiley & Sons: Hoboken, NJ, USA, 2015.
13. Ma, R.; Zhitomirsky, I. Electrophoretic deposition of chitosan–albumin and alginate–albumin films. *Surf. Eng.* **2011**, *27*, 51–56. [CrossRef]
14. Lovsky, Y.; Lewis, A.; Sukenik, C.; Grushka, E. Atomic-force-controlled capillary electrophoretic nanoprinting of proteins. *Anal. Bioanal. Chem.* **2010**, *396*, 133–138. [CrossRef]
15. Huang, W.C.; Hu, S.H.; Liu, K.H.; Chen, S.Y.; Liu, D.M. A flexible drug delivery chip for the magnetically-controlled release of anti-epileptic drugs. *J. Control. Release* **2009**, *139*, 221–228. [CrossRef] [PubMed]
16. Levy, Y.; Tal, N.; Tzemach, G.; Weinberger, J.; Domb, A.J.; Mandler, D. Drug-eluting stent with improved durability and controllability properties, obtained via electrocoated adhesive promotion layer. *J. Biomed. Mater. Res. Part B* **2009**, *91*, 819–830. [CrossRef]
17. Gomez, N.; Schmidt, C.E. Nerve growth factor-immobilized polypyrrole: Bioactive electrically conducting polymer for enhanced neurite extension. *J. Biomed. Mater. Res. Part A* **2007**, *81*, 135–149. [CrossRef]
18. Quigley, A.F.; Bulluss, K.J.; Kyratzis, I.L.B.; Gilmore, K.; Mysore, T.; Schirmer, K.S.U.; Kennedy, E.L.; O'Shea, M.; Truong, Y.B.; Edwards, S.L.; et al. Engineering a multimodal nerve conduit for repair of injured peripheral nerve. *J. Neural Eng.* **2013**, *10*, 016008. [CrossRef] [PubMed]
19. Ducheyne, P.; Van Raemdonck, W.; Heughebaert, J.C.; Heughebaert, M. Structural analysis of hydroxyapatite coatings on titanium. *Biomaterials* **1986**, *7*, 97–103. [CrossRef]
20. Kubie, L.S.; Shults, G.M. Studies on the relationship of the chemical constituents of blood and cerebrospinal fluid. *J. Exp. Med.* **1925**, *42*, 565–591. [CrossRef]
21. Scales, J.T.; Winter, G.D.; Shirley, H.T. Corrosion of orthopaedic implants: Screws, plates and femoral nail-plates. *J. Bone Jt. Surg.* **1959**, *41*, 810–820. [CrossRef]

22. Williams, D.F. Tissue-biomaterial interactions. *J. Mater. Sci.* **1987**, *22*, 3421–3445. [CrossRef]

23. *ASTM F-86 04 Standard Practice for Surface Preparation and Marking of Metallic Surgical Implants*; ASTM International: West Conshohocken, PA, USA, 2004.

24. Hübler, R.; Cozza, A.; Marcondes, T.L.; Souza, R.B.; Fiori, F.F. Wear and corrosion protection of 316-L femoral implants by deposition of thin films. *Surf. Coat. Technol.* **2001**, *142*, 1078–1083. [CrossRef]

25. Lemons, J.; Venugopalan, R.; Lucas, L. Corrosion and biodegradation. In *Handbook of Biomaterials Evaluation: Scientific, Technical, and Clinical Testing of Implant Materials*; von Recum, A.F., Ed.; Taylor & Francis: New York, NY, USA, 1999; pp. 155–167.

26. Olmedo, D.; Fernández, M.M.; Guglielmotti, M.B.; Cabrini, R.L. Macrophages related to dental implant failure. *Implant Dent.* **2003**, *12*, 75–80. [CrossRef]

27. Fleck, C.; Eifler, D. Corrosion, fatigue and corrosion fatigue behaviour of metal implant materials, especially titanium alloys. *Int. J. Fatigue* **2010**, *32*, 929–935. [CrossRef]

28. Eisenbarth, E.; Velten, D.; Müller, M.; Thull, R.; Breme, J. Biocompatibility of β-stabilizing elements of titanium alloys. *Biomaterials* **2004**, *25*, 5705–5713. [CrossRef] [PubMed]

29. Niinomi, M.; Boehlert, C.J. Titanium alloys for biomedical applications. In *Advances in Metallic Biomaterials*; Niinomi, M., Narushima, T., Nakai, M., Eds.; Springer: Berlin, Germany, 2015; pp. 179–213.

30. Walker, P.R.; LeBlanc, J.; Sikorska, M. Effects of aluminum and other cations on the structure of brain and liver chromatin. *Biochemistry* **1989**, *28*, 3911–3915. [CrossRef] [PubMed]

31. Liu, X.; Chu, P.K.; Ding, C. Surface modification of titanium, titanium alloys and related materials for biomedical applications. *Mater. Sci. Eng. R* **2004**, *47*, 49–121. [CrossRef]

32. Mohan, L.; Anandan, C.; Grips, V.W. Corrosion behavior of titanium alloy Beta-21S coated with diamond like carbon in Hank's solution. *Appl. Surf. Sci.* **2012**, *258*, 6331–6340. [CrossRef]

33. Picraux, S.T.; Pope, L.E. Tailored surface modification by ion implantation and laser treatment. *Science* **1984**, *226*, 615–622. [CrossRef]

34. Singh, R.; Tiwari, S.K.; Mishra, S.K.; Dahotre, N.B. Electrochemical and mechanical behavior of laser processed Ti–6Al–4V surface in Ringer's physiological solution. *J. Mater. Sci. Mater. Med.* **2011**, *22*, 1787. [CrossRef] [PubMed]

35. Yue, T.M.; Yu, J.K.; Mei, Z.; Man, H.C. Excimer laser surface treatment of Ti–6Al–4V alloy for corrosion resistance enhancement. *Mater. Lett.* **2002**, *52*, 206–212. [CrossRef]

36. Wang, S.; Liu, Y.; Zhang, C.; Liao, Z.; Liu, W. The improvement of wettability, biotribological behavior and corrosion resistance of titanium alloy pretreated by thermal oxidation. *Tribol. Int.* **2014**, *79*, 174–182. [CrossRef]

37. Richard, C.; Kowandy, C.; Landoulsi, J.; Geetha, M.; Ramasawmy, H. Corrosion and wear behavior of thermally sprayed nano ceramic coatings on commercially pure titanium and Ti–13Nb–13Zr substrates. *Int. J. Refract. Met. Hard Mater* **2010**, *28*, 115–123. [CrossRef]

38. Arslan, E.; Totik, Y.; Demirci, E.; Alsaran, A. Influence of surface roughness on corrosion and tribological behavior of CP-Ti after thermal oxidation treatment. *J. Mater. Eng. Perform.* **2010**, *19*, 428–433. [CrossRef]

39. Zhou, H.; Li, F.; He, B.; Wang, J. Air plasma sprayed thermal barrier coatings on titanium alloy substrates. *Surf. Coat. Technol.* **2007**, *201*, 7360–7367. [CrossRef]

40. Cassar, G.; Wilson, J.A.B.; Banfield, S.; Housden, J.; Matthews, A.; Leyland, A. A study of the reciprocating-sliding wear performance of plasma surface treated titanium alloy. *Wear* **2010**, *269*, 60–70. [CrossRef]

41. Liu, Y.Z.; Zu, X.T.; Wang, L.; Qiu, S.Y. Role of aluminum ion implantation on microstructure, microhardness and corrosion properties of titanium alloy. *Vacuum* **2008**, *83*, 444–447. [CrossRef]

42. Wang, Y.M.; Guo, L.X.; Ouyang, J.H.; Zhou, Y.; Jia, D.C. Interface adhesion properties of functional coatings on titanium alloy formed by microarc oxidation method. *Appl. Surf. Sci.* **2009**, *255*, 6875–6880. [CrossRef]

43. Jiang, X.P.; Wang, X.Y.; Li, J.X.; Li, D.Y.; Man, C.S.; Shepard, M.J.; Zhai, T. Enhancement of fatigue and corrosion properties of pure Ti by sandblasting. *Mater. Sci. Eng. A* **2006**, *429*, 30–35. [CrossRef]

44. Guilherme, A.S.; Henriques, G.E.P.; Zavanelli, R.A.; Mesquita, M.F. Surface roughness and fatigue performance of commercially pure titanium and Ti-6Al-4V alloy after different polishing protocols. *J. Prosthet. Dent.* **2005**, *93*, 378–385. [CrossRef] [PubMed]

45. Arenas, M.A.; Tate, T.J.; Conde, A.; De Damborenea, J. Corrosion behaviour of nitrogen implanted titanium in simulated body fluid. *Br. Corros. J.* **2000**, *35*, 232–236. [CrossRef]

46. Zhou, L.; Lv, G.H.; Ji, C.; Yang, S.Z. Application of plasma polymerized siloxane films for the corrosion protection of titanium alloy. *Thin Solid Films* **2012**, *520*, 2505–2509. [CrossRef]

47. Stanfield, J.R.; Bamberg, S. Durability evaluation of biopolymer coating on titanium alloy substrate. *J. Mech. Behav. Biomed. Mater.* **2014**, *35*, 9–17. [CrossRef] [PubMed]

48. Catauro, M.; Bollino, F.; Giovanardi, R.; Veronesi, P. Modification of Ti6Al4V implant surfaces by biocompatible TiO_2/PCL hybrid layers prepared via sol-gel dip coating: Structural characterization, mechanical and corrosion behavior. *Mater. Sci. Eng. C* **2017**, *74*, 501–507. [CrossRef] [PubMed]

49. Szaraniec, B.; Pielichowska, K.; Pac, E.; Menaszek, E. Multifunctional polymer coatings for titanium implants. *Mater. Sci. Eng. C* **2018**, *93*, 950–957. [CrossRef] [PubMed]

50. DeBerry, D.W. Modification of the electrochemical and corrosion behavior of stainless steels with an electroactive coating. *J. Electrochem. Soc.* **1985**, *132*, 1022–1026. [CrossRef]

51. Mengoli, G.; Munari, M.T.; Bianco, P.; Musiani, M.M. Anodic synthesis of polyaniline coatings onto Fe sheets. *J. Appl. Polym. Sci.* **1981**, *26*, 4247–4257. [CrossRef]

52. Cram, S.L.; Spinks, G.M.; Wallace, G.G.; Brown, H.R. Mechanism of electropolymerisation of methyl methacrylate and glycidyl acrylate on stainless steel. *Electrochim. Acta* **2002**, *47*, 1935–1948. [CrossRef]

53. De Giglio, E.; Guascito, M.R.; Sabbatini, L.; Zambonin, G. Electropolymerization of pyrrole on titanium substrates for the future development of new biocompatible surfaces. *Biomaterials* **2001**, *22*, 2609–2616. [CrossRef]

54. Flamini, D.O.; Saidman, S.B. Electrodeposition of polypyrrole onto NiTi and the corrosion behaviour of the coated alloy. *Corros. Sci.* **2010**, *52*, 229–234. [CrossRef]

55. Mîndroiu, M.; Pirvu, C.; Cimpean, A.; Demetrescu, I. Corrosion and biocompatibility of PPy/PEG coating electrodeposited on Ti6Al7Nb alloy. *Mater. Corros.* **2013**, *64*, 926–931. [CrossRef]

56. Ungureanu, C.; Popescu, S.; Purcel, G.; Tofan, V.; Popescu, M.; Sălăgeanu, A.; Pîrvu, C. Improved antibacterial behavior of titanium surface with torularhodin–polypyrrole film. *Mater. Sci. Eng. C* **2014**, *42*, 726–733. [CrossRef] [PubMed]

57. Mattioli-Belmonte, M.; Gabbanelli, F.; Marcaccio, M.; Giantomassi, F.; Tarsi, R.; Natali, D.; Paolucci, F.; Biagini, G. Bio-characterisation of tosylate-doped polypyrrole films for biomedical applications. *Mater. Sci. Eng. C* **2005**, *25*, 43–49. [CrossRef]

58. De Giglio, E.; Cometa, S.; Sabbatini, L.; Zambonin, P.G.; Spoto, G. Electrosynthesis and analytical characterization of PMMA coatings on titanium substrates as barriers against ion release. *Anal. Bioanal. Chem.* **2005**, *381*, 626–633. [CrossRef] [PubMed]

59. De Giglio, E.; Cometa, S.; Cioffi, N.; Torsi, L.; Sabbatini, L. Analytical investigations of poly (acrylic acid) coatings electrodeposited on titanium-based implants: a versatile approach to biocompatibility enhancement. *Anal. Bioanal. Chem.* **2007**, *389*, 2055–2063. [CrossRef]

60. De Giglio, E.; Cafagna, D.; Ricci, M.A.; Sabbatini, L.; Cometa, S.; Ferretti, C.; Mattioli-Belmonte, M. Biocompatibility of poly (acrylic acid) thin coatings electro-synthesized onto TiAlV-based implants. *J. Bioact. Compat. Polym.* **2010**, *25*, 374–391. [CrossRef]

61. Meng, L.; Li, Y.; Pan, K.; Zhu, Y.; Wei, W.; Li, X.; Liu, X. Colloidal particle based electrodeposition coatings on NiTi alloy: Reduced releasing of nickel ions and improved biocompatibility. *Mater. Lett.* **2018**, *230*, 228–231. [CrossRef]

62. Moskalewicz, T.; Zych, A.; ukaszczyk, A.; Cholewa-Kowalska, K.; Kruk, A.; Dubiel, B.; Radziszewska, A.; Berent, K.; Gajewska, M. Electrophoretic deposition, microstructure and corrosion resistance of porous sol–gel glass/polyetheretherketone coatings on the Ti-13Nb-13Zr alloy. *Metall. Mater. Trans. A* **2017**, *48*, 2660–2673. [CrossRef]

63. Sak, A.; Moskalewicz, T.; Zimowski, S.; Cieniek, .; Dubiel, B.; Radziszewska, A.; Kot, M.; ukaszczyk, A. Influence of polyetheretherketone coatings on the Ti–13Nb–13Zr titanium alloy's bio-tribological properties and corrosion resistance. *Mater. Sci. Eng. C* **2016**, *63*, 52–61. [CrossRef]

64. Jugowiec, D.; ukaszczyk, A.; Cieniek, .; Kot, M.; Reczyńska, K.; Cholewa-Kowalska, K.; Pamuła, E.; Moskalewicz, T. Electrophoretic deposition and characterization of composite chitosan-based coatings incorporating bioglass and sol-gel glass particles on the Ti-13Nb-13Zr alloy. *Surf. Coat. Technol.* **2017**, *319*, 33–46. [CrossRef]

65. Kumar, A.M.; Hussein, M.A.; Adesina, A.Y.; Ramakrishna, S.; Al-Aqeeli, N. Influence of surface treatment on PEDOT coatings: Surface and electrochemical corrosion aspects of newly developed Ti alloy. *RSC Adv.* **2018**, *8*, 19181–19195. [CrossRef]

66. Bosh, N.; Deggelmann, L.; Blattert, C.; Mozaffari, H.; Müller, C. Synthesis and characterization of Halar®polymer coating deposited on titanium substrate by electrophoretic deposition process. *Surf. Coat. Technol.* **2018**, *347*, 369–378. [CrossRef]

67. Manam, N.S.; Harun, W.S.W.; Shri, D.N.A.; Ghani, S.A.C.; Kurniawan, T.; Ismail, M.H.; Ibrahim, M.H.I. Study of corrosion in biocompatible metals for implants: A review. *J. Alloy. Compd.* **2017**, *701*, 698–715. [CrossRef]

68. Ortiz-Hernandez, M.; Rappe, K.; Molmeneu, M.; Mas-Moruno, C.; Guillem-Marti, J.; Punset, M.; Caparros, C.; Calero, J.; Franch, J.; Fernandez-Fairen, M.; et al. Two different strategies to enhance osseointegration in porous titanium: Inorganic thermo-chemical treatment versus organic coating by peptide adsorption. *Int. J. Mol. Sci.* **2018**, *19*, 2574. [CrossRef] [PubMed]

69. Ren, B.; Wan, Y.; Wang, G.; Liu, Z.; Huang, Y.; Wang, H. Morphologically modified surface with hierarchical micro-/nano-structures for enhanced bioactivity of titanium implants. *J. Mater. Sci.* **2018**, *53*, 12679–12691. [CrossRef]

70. Rautray, T.R.; Narayanan, R.; Kwon, T.Y.; Kim, K.H. Surface modification of titanium and titanium alloys by ion implantation. *J. Biomed. Mater. Res. Part B* **2010**, *93*, 581–591. [CrossRef]

71. Zhao, X.; Peng, C.; You, J. Plasma-sprayed ZnO/TiO$_2$ coatings with enhanced biological performance. *J. Therm. Spray Technol.* **2017**, *26*, 1301–1307. [CrossRef]

72. Peng, L.; Zhou, S.; Yang, B.; Bao, M.; Chen, G.; Zhang, X. Chemically modified surface having a dual-structured hierarchical topography for controlled cell growth. *ACS Appl. Mater. Interfaces* **2017**, *9*, 24339–24347. [CrossRef]

73. Liang, J.; Song, R.; Huang, Q.; Yang, Y.; Lin, L.; Zhang, Y.; Jiang, P.; Duan, H.; Dong, X.; Lin, C. Electrochemical construction of a bio-inspired micro/nano-textured structure with cell-sized microhole arrays on biomedical titanium to enhance bioactivity. *Electrochim. Acta* **2015**, *174*, 1149–1159. [CrossRef]

74. Jeon, H.; Simon, C.G., Jr.; Kim, G. A mini-review: Cell response to microscale, nanoscale and hierarchical patterning of surface structure. *J. Biomed. Mater. Res. Part B* **2014**, *102*, 1580–1594. [CrossRef] [PubMed]

75. Hanawa, T. Biofunctionalization of titanium for dental implant. *Jpn. Dent. Sci. Rev.* **2010**, *46*, 93–101. [CrossRef]

76. De Giglio, E.; Cafagna, D.; Giangregorio, M.M.; Domingos, M.; Mattioli-Belmonte, M.; Cometa, S. PHEMA-based thin hydrogel films for biomedical applications. *J. Bioact. Compat. Polym.* **2011**, *26*, 420–434. [CrossRef]

77. Bhattarai, D.P.; Shrestha, S.; Shrestha, B.K.; Park, C.H.; Kim, C.S. A controlled surface geometry of polyaniline doped titania nanotubes biointerface for accelerating MC3T3-E1 cells growth in bone tissue engineering. *Biochem. Eng. J.* **2018**, *350*, 57–68. [CrossRef]

78. Popescu, S.; Ungureanu, C.; Albu, A.M.; Pirvu, C. Poly(dopamine) assisted deposition of adherent PPy film on Ti substrate. *Prog. Org. Coat.* **2014**, *77*, 1890–1900. [CrossRef]

79. Kamata, H.; Suzuki, S.; Tanaka, Y.; Tsutsumi, Y.; Doi, H.; Nomura, N.; Hanawa, T.; Moriyama, K. Effects of pH, potential and deposition time on the durability of collagen electrodeposited to titanium. *Mater. Trans.* **2011**, *52*, 81–89. [CrossRef]

80. Zhuang, J.; Lin, S.; Dong, L.; Cheng, K.; Weng, W. Magnetically assisted electrodeposition of aligned collagen coatings. *ACS Biomater. Sci. Eng.* **2018**, *4*, 1528–1535. [CrossRef]

81. De Giglio, E.; Cometa, S.; Satriano, C.; Sabbatini, L.; Zambonin, P.G. Electrosynthesis of hydrogel films on metal substratesfor the development of coatings with tunable drug delivery performances. *J. Biomed. Mater. Res. Part A* **2008**, *15*, 1048–1057.

82. Gulati, K.; Ramakrishnan, S.; Aw, M.S.; Atkins, G.J.; Findlay, D.M.; Losic, D. Biocompatible polymer coating of titania nanotube arrays for improved drug elution and osteoblast adhesion. *Acta Biomater.* **2011**, *8*, 449–456. [CrossRef] [PubMed]

83. Zhao, P.; Liu, H.; Deng, H.; Xiao, L.; Qin, C.; Du, Y.; Shi, X. A study of chitosan hydrogel with embedded mesoporous silica nanoparticles loaded by ibuprofen as a dual stimuli-responsive drug release system for surface coating of titanium implants. *Colloids Surf. B* **2014**, *123*, 657–663. [CrossRef]

84. De Giglio, E.; Cometa, S.; Ricci, M.A.; Zizzi, A.; Cafagna, D.; Manzotti, S.; Sabbatini, L.; Mattioli-Belmonte, M. Development and characterization of rhVEGF-loaded poly (HEMA–MOEP) coatings electrosynthesized on titanium to enhance bone mineralization and angiogenesis. *Acta Biomater.* **2010**, *6*, 282–290. [CrossRef]

85. Mattioli-Belmonte, M.; Orciani, M.; Ferretti, C.; Orsini, G.; De Giglio, E.; Di Primio, R. Cell behaviour on bioactive polymeric coatings. *Ital. J. Anat. Embryol.* **2010**, *115*, 127–133.

86. De Giglio, E.; Sabbatini, L.; Zambonin, P.G. Development and analytical characterization of cysteine-grafted polypyrrole films electrosynthesized on Ptand Ti-substrates as precursors of bioactive interfaces. *J. Biomater. Sci. Polym. Ed.* **1999**, *10*, 845–858. [CrossRef]

87. De Giglio, E.; Sabbatini, L.; Colucci, S.; Zambonin, G. Synthesis, analytical characterization and osteoblast adhesion properties on RGD-grafted polypyrrole coatings on titanium substrates. *J. Biomater. Sci. Polym. Ed.* **2000**, *11*, 1073–1083. [CrossRef]

88. Buxadera-Palomero, J.; Calvo, C.; Torrent-Camarero, S.; Gil, F.J.; Mas-Moruno, C.; Canal, C.; Rodríguez, D. Biofunctional polyethylene glycol coatings on titanium: An in vitro-based comparison of functionalization methods. *Colloids Surf. B* **2017**, *152*, 367–375. [CrossRef] [PubMed]

89. De Giglio, E.; Cometa, S.; Calvano, C.D.; Sabbatini, L.; Zambonin, P.G.; Colucci, S.; Di Benedetto, A.; Colaianni, G. A new titanium biofunctionalized interface based on poly(pyrrole-3-acetic acid) coating: Proliferation of osteoblast-like cells and future perspectives. *J. Mater. Sci. Mater. Med.* **2007**, *18*, 1781–1789. [CrossRef] [PubMed]

90. Li, G.; Yang, P.; Huang, N. Layer-by-layer construction of the heparin/fibronectin coatings on titanium surface: stability and functionality. *Phys. Procedia* **2011**, *18*, 112–121. [CrossRef]

91. Ye, S.H.; Johnson, C.A., Jr.; Woolley, J.R.; Oh, H.I.; Gamble, L.J.; Ishihara, K.; Wagner, W.R. Surface modification of a titanium alloy with a phospholipid polymer prepared by a plasma-induced grafting technique to improve surface thromboresistance. *Colloids Surf. B* **2009**, *74*, 96–102. [CrossRef] [PubMed]

92. Durairaj, R.B.; Ramachandran, S. In vitro characterization of electrodeposited hydroxyapatite coatings on titanium (Ti6AL4V) and magnesium (AZ31) alloys for biomedical application. *Int. J. Electrochem. Sci.* **2018**, *13*, 4841–4854. [CrossRef]

93. Vranceanu, D.M.; Tran, T.; Ungureanu, E.; Negoiescu, V.; Tarcolea, M.; Dinu, M.; Vladescu, A.; Zamfir, R.; Timotin, A.C.; Cotrut, C.M. Pulsed electrochemical deposition of Ag doped hydroxyapatite bioactive coatings on Ti6Al4V for medical purposes. *Univ. Politeh. Buchar. Sci. Bull. Ser. B Chem. Mater. Sci.* **2018**, *80*, 173–184.

94. Utku, F.S.; Seckin, E.; Goller, G.; Tamerler, C.; Urgen, M. Electrochemically designed interfaces: Hydroxyapatite coated macro-mesoporous titania surfaces. *Appl. Surf. Sci.* **2015**, *350*, 62–68. [CrossRef]

95. De Giglio, E.; De Gennaro, L.; Sabbatini, L.; Zambonin, G. Analytical characterization of collagen-and/or hydroxyapatite-modified polypyrrole films electrosynthesized on Ti-substrates for the development of new bioactive surfaces. *J. Biomater. Sci. Polym. Ed.* **2001**, *12*, 63–76. [CrossRef]

96. Zhang, S.; Cheng, X.; Shi, J.; Pang, J.; Wang, Z.; Shi, W.; Liu, F.; Ji, B. Electrochemical deposition of calcium phosphate/chitosan/gentamicin on a titanium alloy for bone tissue healing. *Int. J. Electrochem. Sci.* **2018**, *13*, 4046–4054. [CrossRef]

97. Avcu, E.; Avcu, Y.Y.; Baştan, F.E.; Rehman, M.A.U.; Üstel, F.; Boccaccini, A.R. Tailoring the surface characteristics of electrophoretically deposited chitosan-based bioactive glass composite coatings on titanium implants via grit blasting. *Prog. Org. Coat.* **2018**, *123*, 362–373. [CrossRef]

98. Mokabber, T.; Lu, L.Q.; Van Rijn, P.; Vakis, A.I.; Pei, Y.T. Crystal growth mechanism of calcium phosphate coatings on titanium by electrochemical deposition. *Surf. Coat. Technol.* **2018**, *334*, 526–535. [CrossRef]

99. Gristina, A.G.; Naylor, P.; Myrvik, Q. Infections from biomaterials and implants: A race for the surface. *Med. Prog. Through Technol.* **1988**, *14*, 205–224.

100. De Giglio, E.; Cometa, S.; Ricci, M.A.; Cafagna, D.; Savino, A.M.; Sabbatini, L.; Orciani, M.; Ceci, E.; Novello, L.; Tantillo, G.M.; et al. Ciprofloxacin-modified electrosynthesized hydrogel coatings to prevent titanium-implant-associated infections. *Acta Biomater.* **2011**, *7*, 882–891. [CrossRef]

101. Cometa, S.; Mattioli-Belmonte, M.; Cafagna, D.; Iatta, R.; Ceci, E.; De Giglio, E. Antibiotic-modified hydrogel coatings on titanium dental implants. *J. Biol. Regul. Homeost. Agents* **2012**, *26*, 65–71. [PubMed]

102. Raj, R.M.; Priya, P.; Raj, V. Gentamicin-loaded ceramic-biopolymer dual layer coatings on the Ti with improved bioactive and corrosion resistance properties for orthopedic applications. *J. Mech. Behav. Biomed. Mater.* **2018**, *82*, 299–309.

103. Hoyos-Nogués, M.; Buxadera-Palomero, J.; Ginebra, M.P.; Manero, J.M.; Gil, F.J.; Mas-Moruno, C. All-in-one trifunctional strategy: A cell adhesive, bacteriostatic and bactericidal coating for titanium implants. *Colloids Surf. B* **2018**, *169*, 30–40. [CrossRef] [PubMed]

104. Vargas-Reus, M.A.; Memarzadeh, K.; Huang, J.; Ren, G.G.; Allaker, R.P. Antimicrobial activity of nanoparticulate metal oxides against peri-implantitis pathogens. *Int. J. Antimicrob. Agents* **2012**, *40*, 135–139. [CrossRef]

105. Memarzadeh, K.; Vargas, M.; Huang, J.; Fan, J.; Allaker, R.P. Nano metallic-oxides as antimicrobials for implant coatings. In *Key Engineering Materials*; Kayali, E.S., Göller, G., Akin, I., Eds.; Trans Tech Publications: Zurich, Switzerland, 2012; Volume 493, pp. 489–494.

106. Sowa-Söhle, E.N.; Schwenke, A.; Wagener, P.; Weiss, A.; Wiegel, H.; Sajti, C.L.; Haverich, A.; Barcikowski, S.; Loos, A. Antimicrobial efficacy, cytotoxicity and ion release of mixed metal (Ag, Cu, Zn, Mg) nanoparticle polymer composite implant material. *BioNanoMaterials* **2013**, *14*, 217–227. [CrossRef]

107. De Giglio, E.; Cafagna, D.; Cometa, S.; Allegretta, A.; Pedico, A.; Giannossa, L.C.; Sabbatini, L.; Mattioli, M.M.; Iatta, R. An innovative, easily fabricated, silver nanoparticle-based titanium implant coating: Development and analytical characterization. *Anal. Bioanal. Chem.* **2013**, *405*, 805–816. [CrossRef] [PubMed]

108. Bonifacio, M.A.; Cometa, S.; Dicarlo, M.; Baruzzi, F.; de Candia, S.; Gloria, A.; Giangregorio, M.M.; Mattioli, M.; De Giglio, E. Gallium-modified chitosan/poly (acrylic acid) bilayer coatings for improved titanium implant performances. *Carbohydr. Polym.* **2017**, *166*, 348–357. [CrossRef]

109. Cometa, S.; Bonifacio, M.A.; Baruzzi, F.; de Candia, S.; Giangregorio, M.M.; Giannossa, L.C.; Dicarlo, M.; Mattioli, M.; Sabbatini, L.; De Giglio, E. Silver-loaded chitosan coating as an integrated approach to face titanium implant-associated infections: analytical characterization and biological activity. *Anal. Bioanal. Chem.* **2017**, *409*, 7211–7221. [CrossRef]

110. Bonifacio, M.A.; Cometa, S.; De Giglio, E. Simulating bacteria-materials interactions via agent-based modeling. In *Italian Workshop on Artificial Life and Evolutionary Computation*; Rossi, F., Mavelli, F., Eds.; Springer: Berlin, Germany, 2015; pp. 77–82.

111. GhavamiNejad, A.; Aguilar, L.E.; Ambade, R.B.; Lee, S.H.; Park, C.H.; Kim, C.S. Immobilization of silver nanoparticles on electropolymerized polydopamine films for metal implant applications. *Colloids Interface Sci. Commun.* **2015**, *6*, 5–8. [CrossRef]

112. Eraković, S.; Janković, A.; Matić, I.Z.; Juranić, Z.D.; Vukašinović-Sekulić, M.; Stevanović, T.; Mišković-Stanković, V. Investigation of silver impact on hydroxyapatite/lignin coatings electrodeposited on titanium. *Mater. Chem. Phys.* **2013**, *142*, 521–530. [CrossRef]

Article

Fungal Growth on Coated Wood Exposed Outdoors: Influence of Coating Pigmentation, Cardinal Direction, and Inclination of Wood Surfaces

Laurence Podgorski *, Céline Reynaud and Mathilde Montibus

FCBA Technological Institute, Allée de Boutaut BP227, Bordeaux F-33028, France; celine.reynaud@fcba.fr (C.R.); mathilde.montibus@fcba.fr (M.M.)
* Correspondence: laurence.podgorski@fcba.fr; Tel.: +33-556-436-366

Received: 21 November 2018; Accepted: 21 December 2018; Published: 4 January 2019

Abstract: Four coating systems were exposed for one year outdoors at 45° south. They consisted of solventborne (alkyd based) and waterborne (acrylic based) systems in both clear and pigmented versions. Fungal growth visually assessed was compared to fungal enumeration, and the influence of exposure time on the main fungal species was studied. Results clearly showed that fungal growth was lower on the pigmented coating systems compared with their pigment-free versions. Although the clear solventborne coating included a higher amount of biocide, it was more susceptible to blue stain than the pigmented version. A new multifaceted exposure rig (MFER) also contributed to the study of fungal growth. It allowed samples to be exposed with nine different exposure directions and angles. Exposure using this MFER has shown that the worst cases (highest area and intensity of blue stain fungi) were for samples with the clear coating system exposed to north 45° and at the top of the MFER (horizontal surfaces). For any cardinal direction, all surfaces inclined at 45° displayed more blue stain fungi than vertical surfaces, due to a higher moisture content of the panels. Depending on the cardinal direction and the orientation, some surfaces were free of visible cracking, but colonized by fungi. It was concluded that the growth of blue stain fungi was not linked with cracking development.

Keywords: coating; wood; weathering; fungi; blue stain; molds; cracking; pigments

---

1. Introduction

Surface fungi or molds grow on most carbon-containing materials including wood, paint, and clear coatings [1–3]. Fungal growth, and especially blue stain fungi on exterior wood coatings are considered to be a major maintenance concern. In addition to reducing the aesthetics of surfaces of buildings, blue stain may also shorten the service life of coatings: when blue stain fungi develop on coated surfaces, the hyphae create pinholes in the coatings, and therefore contribute to coating film disruption [4]. These pinholes are pathways for moisture ingress, leading to the possible decay of building components. Avoiding blue stain is possible by using fungicides in the coating. However, due to the Biocidal Products Regulation, anti-blue stain fungicides are less and less being used in coatings [5,6]. This leads to surface coatings on buildings with a higher amount of fungal growth. Modern heat insulation also contributes to mold growth, as surfaces remain damp for longer periods [6,7]. Several studies have reported that coating formulation and especially pigmentation influence fungal growth [4,8–10]. It was shown that brown semi-transparent coatings were less sensitive to blue stain than white paints [9].

Fungal growth is overlooked in the performance criteria of coating systems (EN 927-2), which is based on the mandatory assessment of blistering, cracking, flaking, and adhesion after 12 months of natural weathering [11]. Blistering is the sign of a lack of water-vapor transmission through the coating film. However, it is rarely noticed with the present waterborne coatings, which have higher moisture permeabilities in comparison to solventborne alternatives [4]. Therefore, amongst these performance criteria, the first sign of coating degradation is cracking. It subsequently leads to flaking and loss of adhesion. According to some authors, blue stain fungi in service are considered not to originate from spores that are already present in the wood, but from penetration through defects in the paint film or insufficiently protected surfaces [4]. However, the relationship between fungal growth and cracking development is still unclear.

Therefore, the objective of this paper was to study the influence of coating pigmentation, cardinal direction, and exposure angle on fungal growth of field-exposed panels, with special attention to the sequence between fungal growth and cracking development. Four coating systems were exposed for one year outdoors at 45° south. They consisted of solventborne (alkyd based) and waterborne (acrylic based) systems, in both clear and pigmented versions. This paper compares visually assessed fungal growth to fungal enumeration. The influence of exposure time on the main fungal species was studied. A new multifaceted exposure rig (MFER) also contributed to the study of fungal growth versus the cardinal orientation and angle of exposure (45°, 90°, 0°).

2. Materials and Methods

2.1. Coatings

Four coatings (two acrylics and two alkyds) described in Table 1 were used. The alkyd coatings (ICP strsp and ICP clear) were based on the Internal Comparison Product (ICP) defined in EN 927-3 [12]. The semi-transparent ICP (ICP strsp) was pigmented, whereas the ICP clear did not include any pigments (Table 2).

Semi-transparent and a clear waterborne formulations, equivalent in pigment volume concentration to the solvent-borne ICP formulations, were designed (WbAcry trsp and WbAcry clear). The acrylic coatings were based on a styrene acrylic emulsion (Table 3).

Table 1. Description of the four tested coatings.

Coating Reference	Type of Coating	Pigmentation
ICP strsp	Solventborne ICP	Semi-transparent
ICP clear	Solventborne ICP	Clear
WbAcry trsp	Waterborne acrylic	Semi-transparent
WbAcry clear	Waterborne acrylic	Clear

Table 2. Description of the solventborne Internal Comparison Product (ICP) in its semi-transparent (ICP strsp) and clear (ICP clear) versions.

Components	ICP Strsp (wt %)	ICP Clear (wt %)
Synolac 6005 WD 65	52.82	52.93
Sicoflush red L2817	4.63	–
Sicoflush yellow L1916	2.30	–
Bentone 34	0.60	0.65
Octa-Solingen Calcium 10	2.77	2.99
Octa-Solingen Cobalt 10	0.37	0.40
Octa-Solingen Zirconium 18	0.30	0.32
Omacide IPBC	0.72	1.00
Tinuvin 292	0.45	0.49
Troysan Anti-skin B	0.20	0.22
Shellsol D40	34.84	41.00
Total	100.00	100.00

Table 3. Description of the waterborne acrylic coatings.

Components	WbAcry Trsp (wt %)	WbAcry Clear (wt %)
DSM Neocryl XK188	86.0	89.7
Water	2.0	2.0
BASF Luconyl 1916 yellow	2.4	–
BASF Luconyl 2817 red	1.2	–
BASF Luconyl 0060 black	0.1	–
Ammonia 25%	0.1	0.1
Ethyldiglycol	5.0	5.0
BASF Lusolvan FBH	1.0	1.0
BASF Dehydran 1293	0.5	0.5
Dow Rocima 250	1.25	1.25
Munzing Tafigel PUR 45	0.45	0.45
Total	100.00	100.00

2.2. Wood Samples

Scots pine sapwood samples fulfilling the requirements of EN 927-3 [12] were selected. They were free from knots, cracks, and resinous streaks. Their dimensions were 375 mm (L), 75 mm (R), 20 mm (T). The mean density was 522 kg/m^3. The inclination of the growth rings to the face was 5° to 45°. They were coated with the four coatings using brushes made of a blend of natural and synthetic bristles (DEXTER). For natural weathering tests (south exposure at 45°), three layers were applied. The wet spreading rate was 50 g/m^2 per layer.

For natural weathering using the multifaceted exposure rig (MFER), the dimensions of samples were 175 mm (L), 75 mm (R), 20 mm (T) with the same wood characteristics as above. Samples were covered with two coatings: ICP clear and WbAcry trsp, both brush-applied in two and three layers, with the same characteristics and spreading rate as before. Uncoated wood samples were also prepared as controls.

2.3. Natural Weathering Test

The samples coated with the four coating systems were exposed for one year (March 2015 to March 2016) at the exposure site of FCBA in Bordeaux (France), on racks facing south and inclined at 45° according to EN 927-3 [12]. For each coating system, three replicates were tested. Uncoated wood samples were also exposed as controls. On these samples, visual assessment of fungal growth as well as fungal enumeration were carried out after 3, 6, 9, and 12 months. Panels were also weighed at each exposure time. The mass obtained was compared with the initial mass conditioned at 20 °C and 65% of relative humidity and corresponding to a theoretical moisture content of 12%.

In addition, a multifaceted exposure rig (MFER) shown in Figure 1 was used. It was made of MONOLUX®500 boards, and was manufactured by PRA (Melton Mowbray, UK) for this study. Its length, width, and height were 1190, 1190, and 1000 mm, respectively. It allowed samples to be exposed to four directions (south, west, north, and east) with two inclinations (45° and 90°). The top of the MFER allowed samples to be exposed horizontally (0°). For each coating system, orientation and inclination, three replicates were used. The samples were exposed for one year (August 2015 to August 2016) at the exposure site of FCBA (Bordeaux, France) and visual assessment of fungal growth was carried out after 3, 6, 9, and 12 months of exposure.

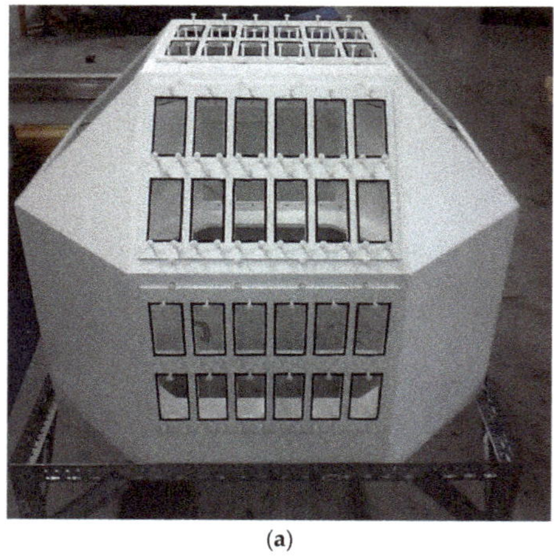

(a)   (b)

Figure 1. The multifaceted exposure rig (MFER) without (**a**) and with samples (**b**).

2.4. Visual Assessment of Fungal Growth

After each exposure time, the fungal growth and the intensity of development were assessed according to EN 16 492 [13].

2.5. Fungal Enumeration on Wood Panels

For fungal analysis, two kinds of enumeration were performed:

- a surface analysis
- an in-depth analysis

For surface analysis, a central surface (50 mm × 40 mm) of each panel was rubbed using a sterile swab moistened with water. The swab was then blended using a Stomacher® 80 (Seward, UK) in 5 mL of sterile water with NaCl 0.9% for 1 min. The recovery solutions were inoculated, sowing 100 µL of serial dilutions on malt/agar. After incubation at 22 °C for 72 h, the enumeration of colonies was undertaken.

For in-depth analysis, the exposed surfaces were cut off from each panel within a thickness of four millimeters; they were sawn, comminuted for 10 s, and then comminuted for another 10 s in 25 mL of sterile water with NaCl 0.9%. The recovery solutions were inoculated, sowing 100 µL of serial dilutions onto malt/agar. After incubation at 22 °C for 72 h, the enumeration of colonies was undertaken. Then, for each sample, the number of colony forming units (CFU) was calculated per cm^2, and the results were then transformed into log_{10} scale (1 LOG (CFU/cm^2) = 10 CFU/cm^2).

Fungal enumeration was performed on samples exposed for 3, 6, 9, and 12 months of south exposure at 45°.

The main fungal species growing after enumeration on panels were isolated for identification.

2.6. Fungal Identification

Fungal identification was achieved using microscopy by mycologists at FCBA. In addition, molecular identification was performed by DNA analysis. DNA extraction was made using the DNeasy plant extraction kit according to the manufacturer's instructions (QIAGEN, Hilden, Germany). PCR amplification was performed using the DreamTaq Hot Start DNA Polymerase, according to the manufacturer's instruction (Thermo Fisher Scientific, Waltham, MA, USA). ITS primers were used as described by Gardes and Bruns [14].

Sequencing was performed by GATC (Konstanz, Germany). Results were compared with the NCBI databases.

2.7. Cracking

The quantity of cracks in the coating systems was assessed after 12 months of exposure using the MFER, and rated from 0 (no detectable cracks) to 5 (dense pattern of cracks) using ISO 4628-4 [15].

3. Results

3.1. Fungal Growth

The area and the intensity of growth versus exposure time are presented in Figures 2 and 3, respectively. The fungal growth was visually identified as a blue stain.

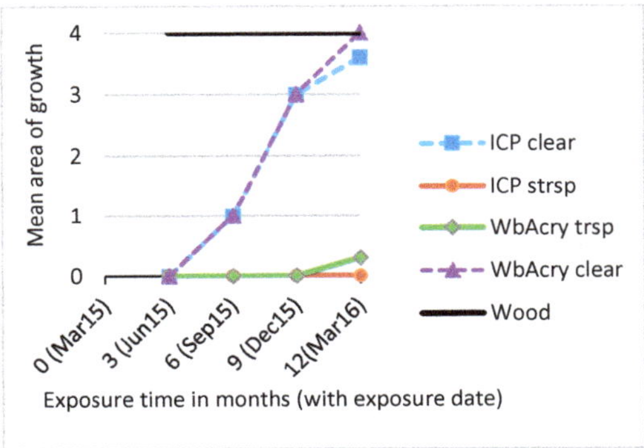

Figure 2. Mean area of fungal growth (EN 16 492) on panels versus exposure time.

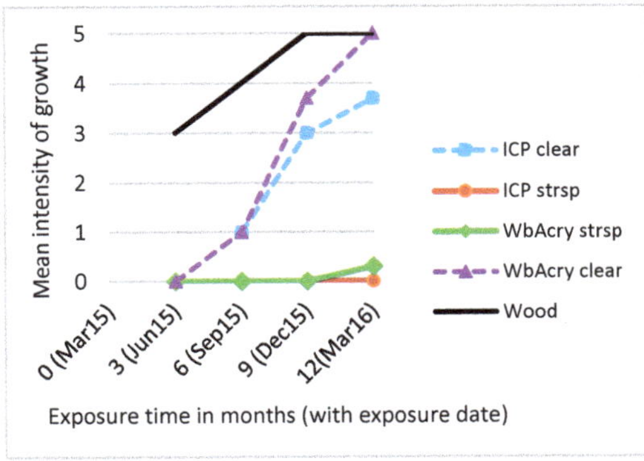

Figure 3. Intensity of fungal growth (EN 16 492) on panels versus exposure time.

Figure 2 shows that for uncoated wood, fungal growth was maximum (rating of 4) starting at 3 months of exposure. For coated wood, a strong difference existed between clear and semi-transparent coating systems. For clear coatings, a rating of 1 was reached after six months of exposure and a rating of 4, similar to wood, was achieved after 12 months of exposure. For pigmented coatings, the rating of 1 was not reached after 12 months of exposure.

Results clearly showed that lower fungal growth was observed on the pigmented coating systems. For the solventborne coatings, the amount of biocide in the recipe was a little bit higher for the clear coating (ICP clear) compared with the pigmented coating (ICP strsp). Despite this higher amount, the clear coating (ICP clear) was more susceptible to blue stain fungi than the pigmented recipe. In the waterborne coatings (WbAcry trsp and WbAcry clear), the amount of biocide was the same, whether the coating was pigmented or not. However, the pigmented waterborne coating system was less prone to blue stain development.

As shown in Figure 3, the intensity of growth for uncoated wood ranged from 3 to 5. For coated wood, a strong difference in the intensity of growth was found between the clear and semi-transparent coating systems. For both clear coating systems, the rating of 1 was reached after six months of exposure. The rating exceeded 3 after 12 months of exposure for the clear ICP, and reached 5 for the clear acrylic coating. For both pigmented coatings (ICP strsp and WbAcry trsp), the rating of 1 was not reached after 12 months of exposure. Results demonstrated that a lower intensity of growth was observed on the pigmented coating systems.

3.2. Fungal Enumeration

The results of fungal enumeration for surface and in-depth analysis are presented in Figures 4 and 5, respectively.

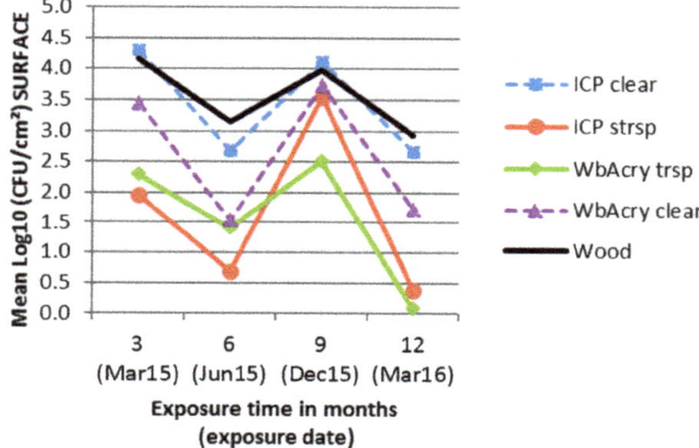

Figure 4. Surface fungal enumeration of panels versus exposure time.

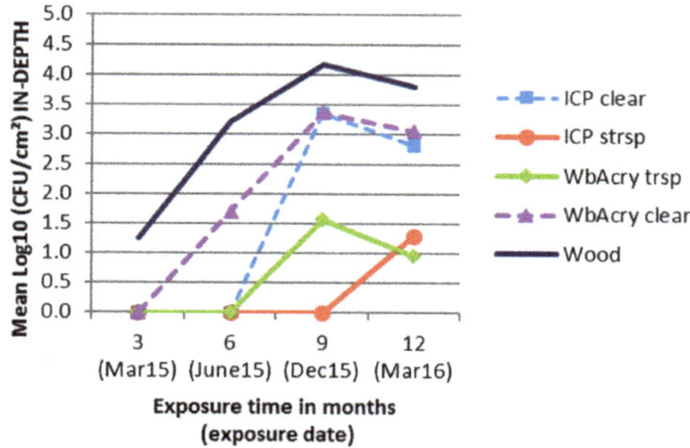

Figure 5. In-depth fungal enumeration of panels versus exposure time.

These figures show that results were different for surface analysis and in-depth analysis. For surface analysis, the time of exposure seemed to influence more the recovering of microorganisms on the surface of panels than the tested coating systems. Between 2 and 4 LOG (CFU/cm^2) were recovered after three and nine months of exposure, whatever the coating system. Between 0 and 3 LOG (CFU/cm^2) were recovered after six and 12 months whatever the coating system. However, the two semi-transparent coating systems were less affected by fungal contamination, which confirmed the visual assessments.

For in-depth analysis, the recovery of microorganisms increased with the time of exposure. After 12 months, the two semi-transparent coating systems were significantly less affected by fungal contamination (1 LOG (CFU/cm^2)) than the clear coating systems (3 LOG (CFU/cm^2)). The results demonstrated that the growth of microorganisms was lower when pigments were present in the coatings.

3.3. Fungal Identification

The main fungi isolated after fungal enumeration were identified to analyze the diversity of microorganisms able to develop on the four tested coating systems after 3, 6, 9, and 12 months of exposure. Results are shown in Figure 6.

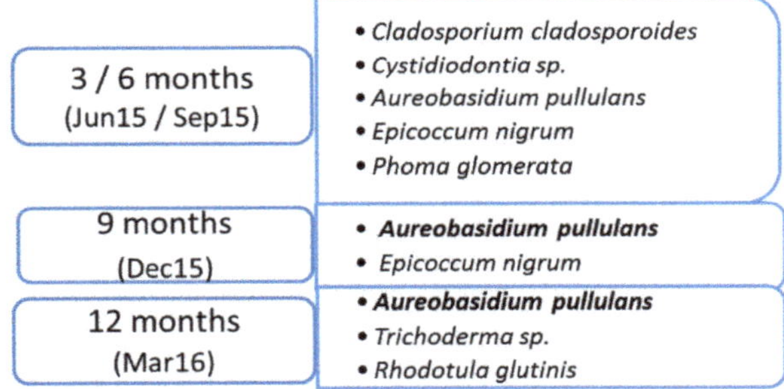

Figure 6. The main fungal species identified after 3, 6, 9, and 12 months of exposure on coated panels (south exposure at 45°). Species in bold were predominantly identified.

After three and six months of exposure, several fungal species were identified, mainly present in the environment. After nine and 12 months of exposure, the diversity of the fungal species significantly decreased, and the blue stain fungus *Aureobasidium pullulans* became dominant. This result also demonstrated that when *Aureobasidium pullulans* becomes dominant, the diversity of other fungal species decreases. This result was consistent with the visual assessments of fungal growth (Figures 2 and 3), which showed an increase in the area and the intensity of fungal growth after nine and 12 months of exposure.

3.4. Influence of Cardinal Direction and Surfaces Inclination on Fungal Growth

Results obtained with samples exposed on the MFER are presented in the following figures.

After three months, only the ICP clear in two and three coats exposed at 45° north displayed some blue stain (mean area = 1). All other surfaces were not disfigured by visible fungal growth.

The results after six months are shown in Figures 7 and 8 for the area and intensity of blue stain, respectively. They show that the pigmented coating systems were less prone to blue stain growth than the clear coating systems. The surfaces inclined at 45° showed more blue stain than the vertical surfaces. Samples exposed to south were less prone to blue stain development.

The results after nine months are shown in Figures 9 and 10 for the area and intensity of blue stain, respectively. The same trends than after six months were observed. For the surfaces inclined at 45°, the lowest development of blue stain was for samples exposed to south.

For some orientations, an additional coat reduced the amount of blue stain. This was clearly the case for the west and east directions. For the north orientation, the results were similar for the two and three coat-applications for both coatings, demonstrating that these surfaces were insufficiently protected. In this case, a fourth coat could probably reduce wetness, and therefore blue stain development.

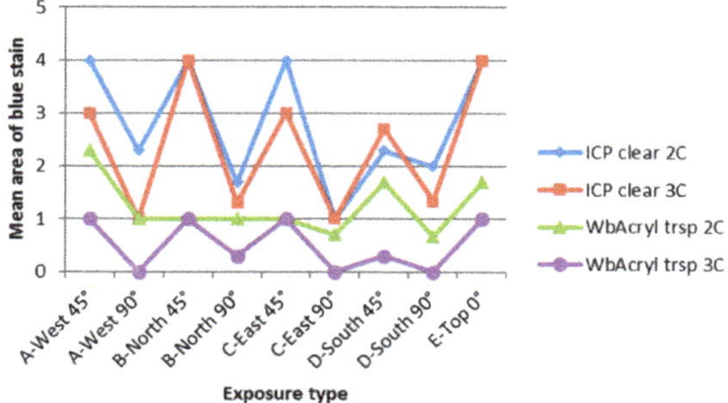

Figure 7. Mean area of blue stain after six months of exposure (February 2016) using the MFER.

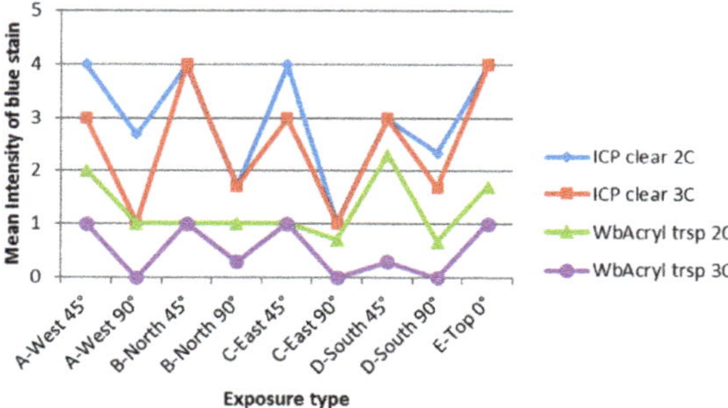

Figure 8. Mean intensity of blue stain after six months of exposure (February 2016) using the MFER.

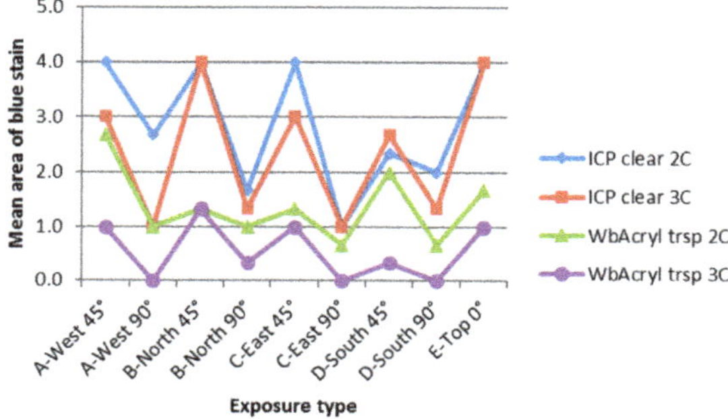

Figure 9. Mean area of blue stain after nine months of exposure (May 2016) using the MFER.

Results after 12 months of exposure are summarized in Figures 11 and 12 for the mean area and the mean intensity of blue stain, respectively.

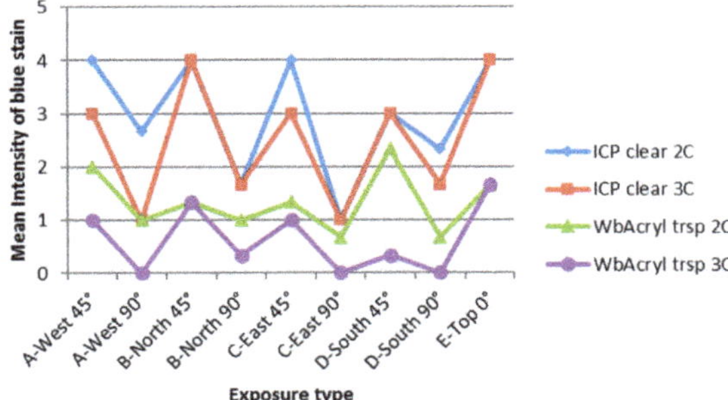

Figure 10. Mean intensity of blue stain after nine months of exposure (May 2016) using the MFER.

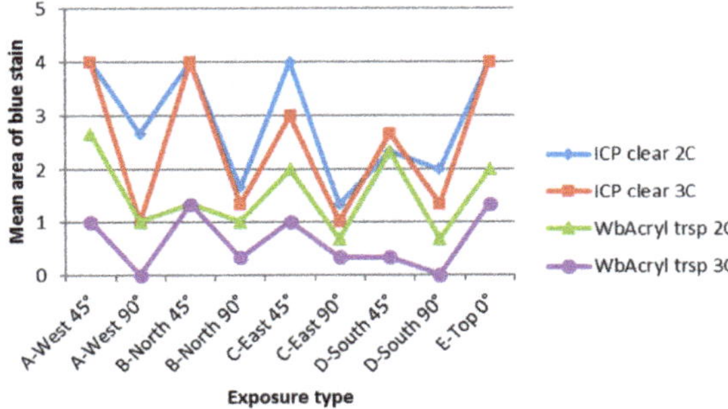

Figure 11. Mean area of blue stain after 12 months of exposure (August 2016) using the MFER.

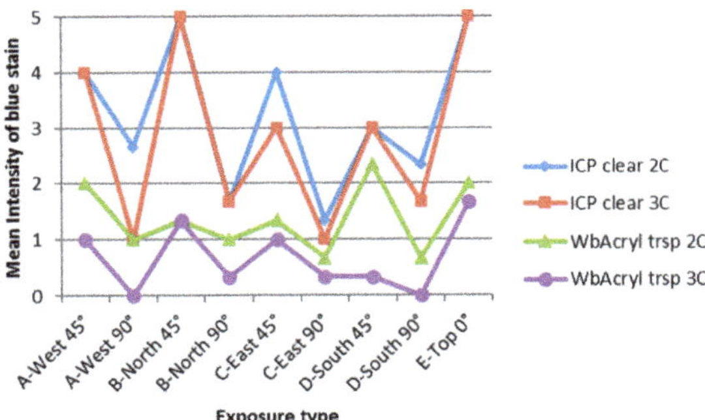

Figure 12. Mean intensity of blue stain after 12 months of exposure (August 2016) using the MFER.

For the clear ICP (two and three coats), the worst cases (highest area and highest intensity) were for samples exposed to north 45° and at the top of the MFER. Samples exposed to west 45° were also prone to a lot of blue stain. For such exposures, results were similar for the two and three coat-applications. The disfigurement of samples depending on orientation and inclination were in good agreement with the variation of the moisture content on the MFER, as shown in Figure 13. This figure also shows that all surfaces inclined at 45° were wetter than those exposed vertically.

For the semi-transparent acrylic coating in two coats, the worst case was for samples exposed at 45° west. An additional coat led to less blue stain for this direction and angle of exposure, demonstrating better protection of surfaces from wetness.

The quantity of cracks in the coating systems, assessed after 12 months of exposure, is shown in Figure 14. It shows that cracking was influenced by cardinal direction and the angle of exposure.

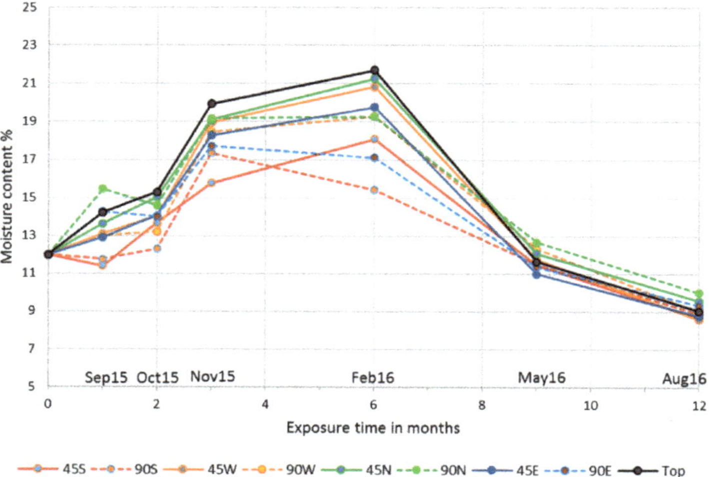

Figure 13. Influence of cardinal direction and the angle of exposure on the moisture content of samples coated with the ICP clear (two coats) exposed on the MFER.

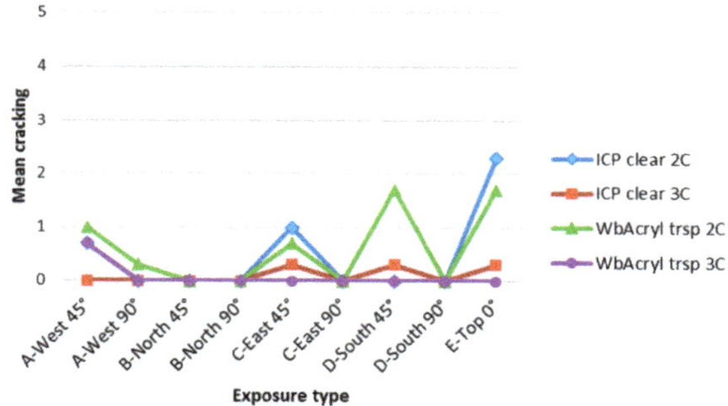

Figure 14. Influence of cardinal direction and the angle of exposure on coating cracking after 12 months of exposure using the MFER.

The comparison of Figure 14 with Figures 11 and 12 (respectively area and intensity of blue stain fungi) clearly shows that the development of blue stain fungi was not linked with the development of cracking. Some surfaces were free of visible cracking, but they displayed significant blue staining. For example, the clear ICP in two coats had no cracking for surfaces exposed to north 45° which displayed the highest ratings for both the area and intensity of blue stain fungi. The clear ICP in two coats displayed cracking for surfaces exposed to west 45°, east 45°, and at the top of the MFER. Cracking was not observed on the vertical surfaces; however, blue stain fungi were observed.

The increase in coating thickness due to a third coat clearly reduced the cracking score of the clear ICP. For the surfaces exposed horizontally, cracking was reduced from 2.3 to 0.3, due to the additional layer of the ICP clear.

For the semi-transparent acrylic coating in two coats, the highest cracking score of 1.7 was for surfaces exposed to south 45° and horizontally. Surfaces exposed to north did not show any visible cracking, but had some blue stain fungi. The addition of a third coat clearly led to a decrease in cracking development. All surfaces, except those exposed to west 45°, were free of cracking, but most of them displayed some blue stain fungi (Figures 11 and 12).

4. Discussion

Results have shown that fungal growth on coatings exposed for one year mainly consisted of blue stain fungi, as was already reported by some authors [4,7,16]. Fungal growth was significantly influenced by coating pigmentation, which confirms the findings of other authors [4,8–10]. Significantly lower fungal growth was observed on the pigment-containing coatings, compared with their pigment-free versions. Despite the clear solventborne coating (ICP clear) including a higher amount of biocide, it was more susceptible to blue stain than the pigmented recipe (ICP strsp). In the waterborne coatings, the amount of biocide was the same, whether the coating was pigmented or not. However, the pigmented waterborne coating was less prone to blue stain development. Several phenomena can contribute to this effect. Pigments lead to an increase in surface temperature, which decreases the time of the wetness of the samples. Furthermore, pigments also protect biocides from UV degradation [17], which result in better protection from fungal growth for pigmented coatings. The pigments included in coatings were iron oxides. Iron has shown to have a toxic effect on some fungi [18,19]. Therefore, iron oxide pigments may have a biocidal effect on blue stain fungi. Pigmented and clear coatings could also display differences in surface acidity, which could contribute to the lower fungal growth on pigmented coatings.

The results of fungal enumeration were clearly different for surface and in-depth analysis of samples (Figures 4 and 5). At sample surfaces, the number of colonies versus exposure time was in good agreement with seasonal fluctuations: it was higher in winter for the wetter months and lower during summer. The seasonal influence on the number of colonies within the first four millimeters of the samples was less clear. Kržišnik et al. reported seasonal variation in color changes of uncoated wood for samples exposed outdoors in Slovenia for four years [20]. According to the authors, the most important reason for seasonal fluctuations was related to fungal melanin, with its formation on one hand being countered by bleaching on the other. In our study, uncoated wood samples, as well as samples covered with the clear coatings showed a gradual increase in the intensity of blue staining over the year (Figure 3). The intensity of fungal growth (EN 16 492) was directly related to color changes. The seasonal fluctuations of color changes described by Kržišnik et al. were therefore not visible in our results.

Solar radiation is one of the main agents causing the degradation of organic materials exposed outdoors. Therefore, weathering studies usually focus on maximizing solar radiation. Test panels were exposed facing the equator (i.e., facing south in the northern hemisphere), at an angle to the horizontal equal to the latitude of the site where they are exposed, receive the greatest total solar energy of any possible fixed orientations [21]. Practical considerations, such as having similar water run-off from specimens in different localities, have often resulted in exposure racks facing the equator at an angle of 45° to the horizontal, regardless of the latitude of the locality [21]. As the latitude of Bordeaux is 45°, the maximum dose of solar radiation was received for samples facing south at 45°. This orientation and inclination clearly underestimated the issue of fungal growth on coated wood. The MFER demonstrated a wider range of natural weathering behavior than traditional exposure racks inclined at 45° facing south. It has been shown that the worst cases (highest area and the intensity of blue stain fungi) were for samples with the clear coating exposed to north 45°, and at the top of the MFER. For such exposures, results were similar between the two and three coats applications demonstrating insufficient coating thickness for these surfaces. These orientations and inclinations also led to the highest moisture content in the samples. A study that compared two sites in Norway showed that highest mold rating was detected facing north at one location, and facing south at the other. The authors concluded that the effect of cardinal direction on mold growth can vary from site to site [22]. This finding shows that the prevailing rain should be taken into consideration when orienting exposure racks for trials related to fungal growth. Wind driven-rain will increase surface wetness after heavy precipitation, and therefore encourage fungal growth.

Cracking of coatings is an indication they lack sufficient flexibility to accommodate the surface strains that develop when wood swells and shrinks [1]. Cracking is often selected to indicate the end of the service life of coatings, and the need for maintenance [23,24]. Coating cracking will lead to higher fluctuations of wood moisture content, especially during winter [25], which encourages colonization by staining fungi. Our results have shown that the growth of blue stain fungi was not due to cracking development, and could start before visible cracking was noticed. This finding is noteworthy, and shows that special attention should be given to the assessment of blue stain fungi during weathering trials.

5. Conclusions

Results have shown that fungal growth on coatings exposed outdoors mainly consisted of blue stain fungi. Fungal growth was significantly influenced by coating pigmentation, cardinal direction, and the angle of exposure.

This study has clarified the sequence between fungal growth and cracking development. It was shown that blue stain fungi started to grow before the cracking of coatings was noticed. The hyphae of such fungi can lead to pinholes forming in the coating system. Therefore, the service life of coating systems may be dramatically shortened because pinholes can increase water uptake of wood samples. This will lead to higher fluctuations in wood moisture content, dimensional variations of the substrate and coating cracking. We conclude that the mandatory performance criteria of coatings described in EN 927-2 should include fungal growth as well.

Coatings manufacturers commonly use weathering trials to develop new products. In the northern hemisphere, racks facing south and inclined at 45° are used. This orientation clearly underestimates the issue of fungal growth. Surfaces exposed to north and inclined at 45° displayed more blue stain, and provided additional information on the coating service life. Dry-film biocides are used to make the coating system durable. In Europe, as a result of the Biocidal Products Regulation, the number of biocides used to protect dry-films has declined. Furthermore, the development of new active substances has almost completely stopped, due to the cost and complexity of registering substances. Therefore, there is a strong need for alternatives and innovation in dry-film protection coating systems. Any means of reducing the surface wetness should be considered when developing coatings. Coating pigmentation modifies the surface temperature, and therefore decreases surface wetness and fungal growth. Increasing the coating thickness with the application of additional layers will result in the better protection of surfaces exposed to critical cardinal directions. Any additives that can reduce moisture in the coating film could also contribute to reduce fungal growth on coated wood.

Author Contributions: Conceptualization, L.P.; Methodology, L.P. and M.M.; Validation, L.P. and M.M.; Formal Analysis, L.P. and M.M.; Investigation, L.P., C.R. and M.M.; Resources, C.R. and M.M.; Data Curation, L.P.; Writing-Original Draft Preparation, L.P. and M.M.; Writing-Review & Editing, L.P.; Visualization, L.P.; Supervision, L.P.; Project Administration, L.P.; Funding Acquisition, L.P.

Funding: This research was funded within the project SERVOWOOD by the Seventh Framework Programme (FP7/2007-2013) of the European Commission (No. FP7-SME-2013-606576).

Acknowledgments: Contributions to the project from all consortium members are acknowledged. Special thanks to Teknos Drywood B.V. (The Netherlands) for providing the two acrylic coatings, and to the Paint Research Association (United Kingdom) for the alkyd coatings.

Conflicts of Interest: The authors declare no conflict of interest.

References

1. Evans, P.D.; Haase, J.G.; Shakri, A.; Seman, B.M.; Kiguchi, M. The search for durable exterior clear coatings for wood. *Coatings* **2015**, *5*, 830–864. [CrossRef]
2. Jellison, J.; Goodell, B.; Daniel, G. The biology and microscopy of building molds: Medical and molecular aspects. In *Development of Commercial Wood Preservatives: Environmental, and Health Issues*; Schultz, T.P., Militz, H., Freeman, M.H., Goodell, B., Nicholas, D.D., Eds.; American Chemical Society: Washington, DC, USA, 2008.
3. Cogulet, A.; Blanchet, P.; Landry, V.; Morris, P. Weathering of wood coated with semi-clear coating: Study of interactions between photo and biodegradation. *Int. Biodeterior. Biodegrad.* **2018**, *129*, 33–41. [CrossRef]
4. De Meijer, M. Review on the durability of exterior wood coatings with reduced VOC-content. *Prog. Org. Coat.* **2001**, *43*, 217–225. [CrossRef]
5. Regulation (EU) No 528/2012 of the European Parliament and of the Council of 22 May 2012 concerning the making available on the market and use of biocidal products. *Off. J. Eur. Union* **2012**, *L167*, 1–123.
6. Jensen, H.; Sandve, M.; Lystvet, S.M. Exterior paint for the future—Will there be any dry-film preservatives left? In Proceedings of the International Research Group on Wood Protection, 48th Annual Meeting, Ghent, Belgium, 4–8 June 2017.
7. Gaylarde, C.G.; Morton, L.H.G.; Loh, K.; Shirakawa, M.A. Biodeterioration of external architectural paint films—A review. *Int. Biodeterior. Biodegrad.* **2011**, *65*, 1189–1198. [CrossRef]
8. Gobakken, L.R.; Vestøl, G.I. Mould growth on spruce claddings and the effect of selected influencing factors after 4 years of outdoor testing. In Proceedings of the International Research Group on Wood Protection, 43rd Annual Meeting, Kuala Lumpur, Malaysia, 6–10 May 2015.
9. Gobakken, L.R.; Westin, M. Surface mould growth on five modified wood substrates coated with three different coating systems when exposed outdoors. *Int. Biodeterior. Biodegrad.* **2008**, *62*, 397–402. [CrossRef]
10. Van Acker, J.; Stevens, M.; Brauwers, C.; Rijckaert, V.; Mol, E. Blue stain resistance of exterior wood coatings as a function of their typology. In Proceedings of the International Research Group on Wood Protection, 29th Annual Meeting, Maastricht, The Netherlands, 14–19 June 1998.
11. *EN 927–2 Paint and Varnishes—Coating Materials and Coating Systems for Exterior Wood—Part 2: Performance Specification*; European Committee for Standardization: Brussels, Belgium, 2014.
12. *EN 927–3 Paint and Varnishes—Coating Materials and Coating Systems for Exterior Wood—Part 3: Natural Weathering Test*; European Committee for Standardization: Brussels, Belgium, 2012.
13. *EN 16 492 Paint and Varnishes—Evaluation of the Surface Disfigurement Caused by Fungi and Algae on Coatings*; European Committee for Standardization: Brussels, Belgium, 2014.
14. Gardes, M.; Bruns, T.D. ITS primers with enhanced specificity for basidiomycetes-application to the identification of mycorrhizae and rusts. *Mol. Ecol.* **1993**, *2*, 113–118. [CrossRef] [PubMed]
15. *EN ISO 4628-4 Paints and Varnishes—Evaluation of Degradation of Coatings—Designation of Quantity and Size of Defects, and of Intensity of Uniform Changes in Appearance—Part 4: Assessment of Degree of Cracking*; European Committee for Standardization: Brussels, Belgium, 2003.
16. Hansen, K. Molds and moldicide formulations for exterior paints and coatings. In *Development of Commercial Wood Preservatives: Environmental, and Health Issues*; Schultz, T.P., Militz, H., Freeman, M.H., Goodell, B., Nicholas, D.D., Eds.; ACS Symposium Series; American Chemical Society: Washington, DC, USA, 2008.
17. Urbanczyk, M.M.; Bester, K.; Borho, N.; Schoknecht, U.; Bollmann, U.E. Influence of pigments on phototransformation of biocides in paints. *J. Hazard. Mater.* **2019**, *364*, 125–133. [CrossRef] [PubMed]
18. Anahid, S.; Yaghmaei, S.; Ghobadinejad, Z. Heavy metal tolerance of fungi. *Sci. Iran.* **2011**, *18*, 502–508. [CrossRef]
19. Wiśnicka, R.; Krzepiłko, A.; Wawryn, J.; Biliński, T. Iron toxicity in yeast. *Acta Microbiol. Pol.* **1997**, *46*, 339–347. [PubMed]
20. Kržišnik, D.; Boštjan, L.; Thaler, N.; Humar, M. Influence of natural and artificial weathering on the colour change of different wood and wood-based materials. *Forests* **2018**, *9*, 488. [CrossRef]
21. Ballantyne, E.R. The effect of orientation and latitude on the solar radiation received by test panels and fences during weathering studies. *Build. Sci.* **1974**, *9*, 191–196. [CrossRef]

22. Gobakken, L.R.; Vestøl, G.I. Effects of microclimate, wood temperature and surface colour on fungal disfigurement on wooden claddings. In Proceedings of the International Research Group on Wood Protection, 43rd Annual Meeting, Kuala Lumpur, Malaysia, 6–10 May 2012.
23. Forsthuber, B.; Grüll, G.; Arnold, M.; Podgorski, L.; Bulian, F. Service life prediction of exterior wood coatings. In Proceedings of the 5th International Conference on Processing Technologies for the Forest and Bio-based Products Industries, Freising, Germany, 20–21 September 2018.
24. Grüll, G.; Truskaller, M.; Podgorski, L.; Bollmus, S.; Tscherne, F. Maintenance procedures and definition of limit states for exterior wood coatings. *Eur. J. Wood Wood Prod.* **2011**, *69*, 443–450. [CrossRef]
25. Grüll, G.; Truskaller, M.; Podgorski, L.; Bollmus, S.; De Windt, I.; Suttie, E. Moisture conditions in coated wood panels during 24 months natural weathering at five sites in Europe. *Wood Mater. Sci. Eng.* **2013**, *8*, 95–110. [CrossRef]

MDPI

St. Alban-Anlage 66

4052 Basel

Switzerland

Tel. +41 61 683 77 34

Fax +41 61 302 89 18

www.mdpi.com

Coatings Editorial Office

E-mail: coatings@mdpi.com

www.mdpi.com/journal/coatings